MW01323641

ELECTRICIAN'S POCKET REFERENCE

ELECTRICIAN'S POCKET REFERENCE

John Traister

McGraw-Hill

New York San Francisco Washington, D.C. Auckland Bogotá
Caracas Lisbon London Madrid Mexico City Milan
Montreal New Delhi San Juan Singapore
Sydney Tokyo Toronto

McGraw-Hill

A Division of The ***McGraw-Hill*** *Companies*

1 2 3 4 5 6 7 8 9 0 DOC/DOC 9 0 1 0 9 8 7

ISBN-0-07-065337-2

Printed and bound by R.R. Donnelley/Crawfordsville.

This book is printed on acid-free paper.

Contents

Preface . . . . *vii*

Chapter 1 Introduction 1

Chapter 2 Codes and Standards . . . 19

Chapter 3 Tools of the Trade 51

Chapter 4 OSHA and Electrical Safety 75

Chapter 5 Raceway Installations . . . 101

Chapter 6 Outlet, Junction, and Pull Boxes 193

Chapter 7 Wiring Methods 231

Chapter 8 Wiring Devices 269

Chapter 9 Panelboards and Switchgear 287

Chapter 10 Conductor Terminations . 307

Chapter 11 Electric Motors 321

Chapter 12 Motor Controls 331

Chapter 13 Lighting Installations . . . 341

Index 363

Preface

Texts used by electricians have normally included page after page of electrical theory, ac and dc circuits, mathematical calculations, and equations — with very little material covering practical applications. *Electrician's Pocket Reference* is designed to change this situation; that is, with the scores of texts already available covering theory, this author chose to concentrate mainly on the practical data and applications, along with the proper attitude one needs in the daily routine of being an electrician.

This book begins with an overview of the electrical construction industry and apprenticeship, touches briefly on essentials such as codes, standards, and print reading, then quickly jumps into practical categories such as tools, electrical materials, equipment, and the implementation of actual installations.

To keep this book pocket-sized, the treatment of some subjects has necessarily been brief, but subjects of greater importance have been dealt with more fully. The various tables and charts have been selected with great care, and only those that are most likely to be consulted by the apprentice electrician on a daily basis have been included. The numerous rules and equations are stated as simply and concisely as possible, and their applications are clearly illustrated by the full solution of many examples.

Care has been taken to arrange the several chapters in a convenient and logical manner, and a very full index increases further the facility with which any given subject may be located.

Regardless of the project — residential, commercial, industrial, or institutional — tables, charts, installation details and other useful data can be found in this concise handbook to enable the work to go more smoothly. Furthermore, solutions to many everyday electrical problems may be found practically at a glance, reducing the normal time required considerably.

Electricians and apprentice electricians cannot afford to be without this book. Furthermore, its usefulness will not end when he or she passes their electrician's exam and becomes a journey worker, because the numerous tables and charts offer information used in the daily routine that cannot possibly be remembered.

The author is indeed grateful to the dozens of manufacturers who supplied reference material and other data so helpful to the task of compiling this handbook.

John Traister
Bentonville, Virginia
1997

ELECTRICIAN'S POCKET REFERENCE

Chapter 1

INTRODUCTION

The construction industry has become the largest industry in the United States — producing more revenue than both the steel and automobile industries combined. In the coming years, hundreds of billions of dollars will be spent on the construction and maintenance of residential, commercial, institutional, and industrial facilities. More than one trillion dollars will be spent on construction materials, labor, and services.

History of the Building Construction Industry

In almost all instances, when an owner decides to have a building constructed, an architect is hired to prepare the complete working drawings and specifications for the building. The drawings usually include the following:

1. A plot plan indicating the location of the building on the property (Figure 1-1).

2. Elevations of all exterior faces of the building (Figures 1-2 and 1-3).

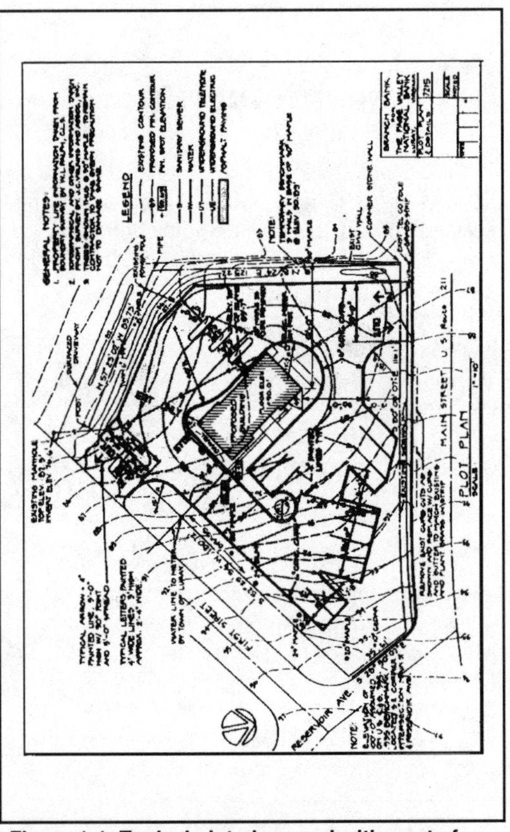

Figure 1-1: Typical plot plan used with a set of architectural drawings.

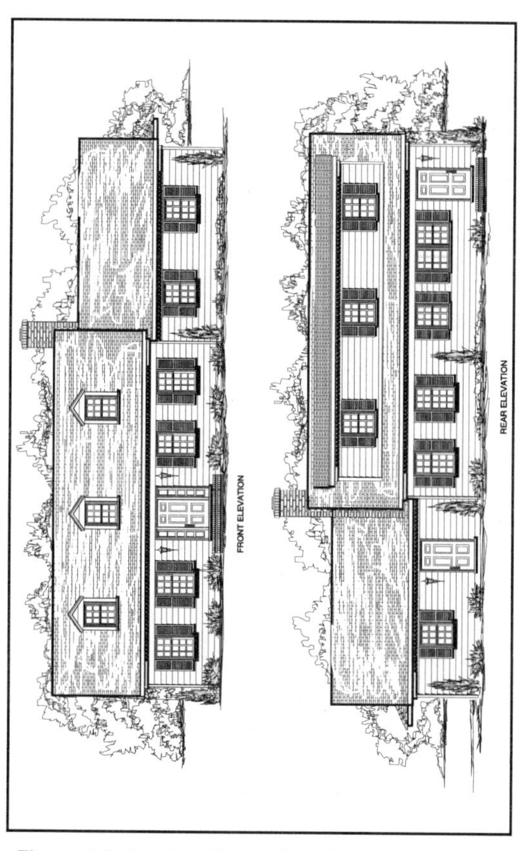

Figure 1-2: Front and rear elevations of a building.

Figure 1-3: Right and left elevations of a building.

3. Floor plans showing the walls and partitions for each floor or level (Figures 1-4 and 1-5).

4. Sufficient vertical cross sections to indicate clearly the various floor levels and details of the foundation, walls, floors, ceilings, and roof construction.

5. Large-scale renderings showing such details of construction as may be required.

The architect often represents the owner in soliciting quotations from general contractors and advises the owner as to the proper award to make. The architect also usually represents the owner during construction of the building and inspects the work to ascertain that it is being performed in accordance with the requirements of the construction documents (drawings and specifications).

On very small projects, the architect may merely show a sketchy diagram of the electrical outlets on the architectural floor plans with switch locations in each area. The wiring is often sized, in cases like this, by the electrician installing the job. For jobs of any consequence, the architect usually includes drawings and specifications covering the design of plumbing, heating, ventilating, air conditioning, and electrical work.

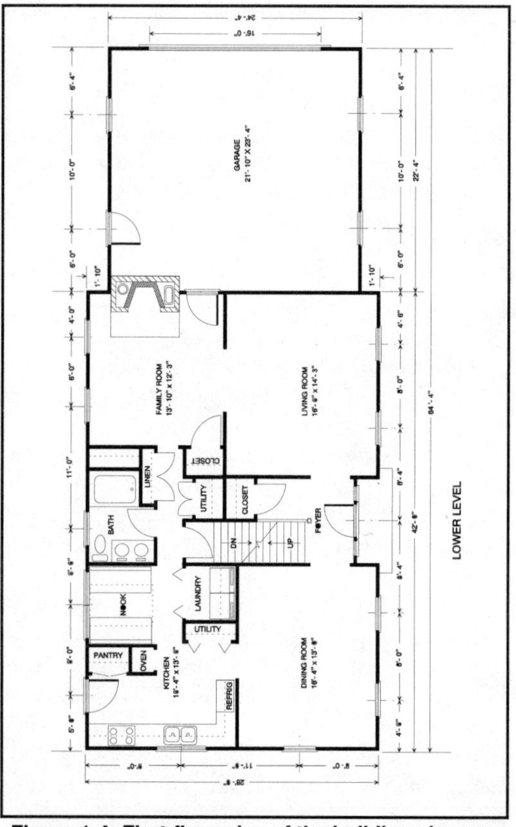

Figure 1-4: First-floor plan of the building shown in Figures 1-2 and 1-3.

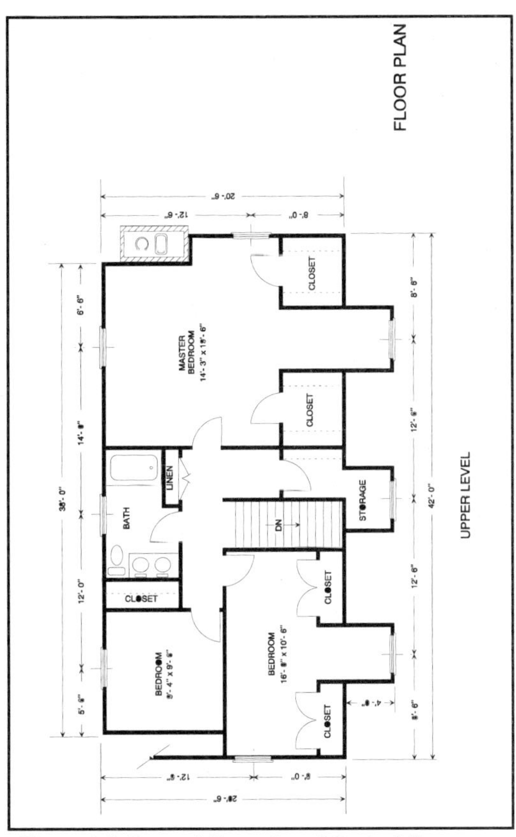

Figure 1-5: Second-floor plan of the building shown in Figures 1-2 and 1-3.

In larger buildings, where the electrical systems are more extensive and complex, architects normally hire consulting engineers to handle the details of the electrical construction from the time the design and layout of the work are started, through the bidding and construction sequences, to the final approval and acceptance of the finished job.

The relations of the consulting engineer parallel those of the architect; that is, the consulting engineer often represents the architect and owner in soliciting quotations from electrical and mechanical contractors. The engineer also inspects his or her portion of the work to assure the architect that this portion is carried out according to the working drawings and specifications. The engineer also approves shop drawings (material submittals), checks and approves progress payments, and performs similar duties — all pertaining only to the engineer's phase of the construction.

Architects sometimes require the engineer to prepare an approximate cost estimate of the mechanical work to aid them in determining the probable cost of the building prior to the actual request for formal quotations. This is especially true when government projects are being developed.

The electrical engineer or designer uses the architect's drawings as a reference when designing a suitable electrical system for the building in question.

This design usually involves calculating the required illumination level in each area of the building, the required circuits and overcurrent protection, the required service size, the size of feeders, panels, switchgear, etc.

Drafters, in turn, are responsible for translating the engineer's design into neat, detailed, and accurate working drawings that indicate, beyond any question of a doubt, exactly what is required to install a correct electrical system in a building. In doing so, the drafter uses lines, symbols, dimensions, and notations to convey the engineer's design to the workers on the job. Usually, a set of written specifications describing in detail the requirements of the design accompanies these drawings. Together with the working drawings, it forms the basis of the contract requirements for the construction of the project.

As an electrician, you will be one of the most important workers in the construction industry. And you may perform any of a wide variety of jobs, depending upon where you work. But basically, the primary function of an electrician is to install or maintain wiring and electrical equipment.

As an electrician, you will work in new houses, office buildings, or factories. You will install new wiring as well as all of the related components, such as breaker boxes, switches, light fixtures, and even telephone and TV wiring. In some cases, you will also install equipment such as motors, machines, and other heavy equipment.

You will also perform maintenance and repairs, replace defective parts as problems occur, and repair equipment when it fails.

Apprenticeship

Classroom training and on-site experience go hand in hand. The beginner — the apprentice — needs both to have a rewarding career.

History of Apprenticeship

Apprenticeship training transfers practical knowledge from one generation to another, and this method of learning a trade goes far back in human history. Provisions for teaching apprentices are found in the Babylonian Code of Hammurabi, which is 4,000 years old. In ancient Rome, the high level of technical achievement that produced aqueducts, great public buildings like the Coliseum, and a large network of roads was due to craftspeople and artisans whose skills were handed down from father to son.

One ancient contract provided that a man named Heraclas be taught weaving during a period of five years. He was given food, 20 annual holidays, and a new tunic each year. After two and a half years, he was paid 12 drachmas; in his fifth year he made twice as much.

During the Middle Ages (500 A.D. to about 1500 A.D.), two social classes developed; merchants and skilled artisans. Both organized themselves into guilds in order to reduce competition and provide training for apprentices. In these training programs, apprentices were indentured (bound by contract) to a master craftsperson who agreed to train them in the craft and provide them with food, clothing, and shelter. Apprenticeships lasted from 2 to 10 years, depending on the trade.

Modern Apprenticeship Training

Apprenticeships remained virtually unchanged until the education reforms of the 19th century, when industrial education became a formal part of the public school system. Compensation changed from food, clothing, and shelter to payment of wages.

Apprenticeship systems gradually developed for a host of new industries that rely on machinery and electrical equipment. The graduated wage scale first appeared in the mid-1800s. In 1911, Wisconsin enacted the first legislation in the United States to regulate apprenticeship systems. This law placed apprenticeships under the authority of an industrial commission.

In the 1920s, national employer and labor organizations, educators, and government officials began an effort to initiate a national apprenticeship system. The construction industry was a prime force in this movement. The need for comprehensive training increased after World War I, when immigration restrictions slowed the flow of skilled foreign workers.

In 1934, Congress passed the Fitzgerald Act — the first step in forming a national system of apprenticeship training. The act established the Bureau of Apprenticeship and Training (BAT), which has since worked closely with employers and labor groups, inspection departments, colleges, vocational schools, and state agencies to promote apprenticeship programs and advise these groups about such training.

Programs registered with BAT must meet the following provisions:

- Apprenticeship opportunities must be available to all.
- Training must be combined with on-the-job experience.
- A minimum of 144 hours per year of related training must be provided.
- The apprenticeship must offer an increasing schedule of wages.
- Apprentices must have proper supervision in adequate facilities.

- Job performance and related instruction must be periodically evaluated.
- Apprentices who successfully complete the program must be formally recognized.

The apprenticeship system is still changing. Today, apprenticeships respond to technological and social conditions brought about by scientific discoveries, new teaching methods, expanding industry, and a booming population. As a result, apprenticeships have been set up in new trades and updated in many older ones.

Advantages of Apprenticeship

Apprenticeship is an efficient way for workers to learn skills because the training is planned and organized. Also, apprentices earn as they learn.

Industry benefits as well. From apprenticeship programs come craft workers who are competent in all aspects of their trade and are able to work independently. When changes are made in production, these workers are able to change work components quickly to suit the changing needs. Quality workers with these skills are vital to industrial progress.

Apprenticeship training provides a ready source of skilled workers to meet current and future employment needs. The on-the-job training component provides the practical skills for the trade. The classroom instruction provides the background information required to understand the trade — what it involves and how the job gets done.

Trade workers who advance in knowledge and skill stay productive and contribute to the economic success of the firm for which they work. Individual initiative is doubly important: workers move up the ranks to positions of greater responsibility and the company remains competitive.

Apprentice Electricians

Electricians work throughout the construction industry, providing electrical services in residential, commercial, and industrial projects. Electricians have one of the highest employment rates of all craft workers. Electricians must be licensed in most areas of the United States and must pass a written exam. All electrical installations must comply with the *National Electrical Code (NEC)*® and the written exam asks many questions that deal with this code. Copies of the code may be obtained from most electrical inspection offices or directly from the National Fire Protection Association, Batterymarch Park, Quincy, MA 02269.

Electricians are frequently exposed to many hazardous conditions and situations. However, they can avoid serious accidents or injuries by being aware of potential hazards and staying constantly alert to them.

The electrician must plan and lay out electrical systems around the work of other trades. Often the requirements for electrical installations are not spelled out in the project's plans and written specifications; therefore, the electrician must be able to develop electrical plans to meet the needs of the consumer. When the requirements are included in the specifications, electricians must make sure that the plans are current.

Electricians must be carefully trained. State-approved apprenticeship programs provide a combination of classroom and on-the-job experience in which trainees work with other skilled electricians. These programs are by far the best way to learn the trade and to keep abreast of *NEC* requirements and changes.

They provide electricians with a foundation that builds on the *NEC*-approved wiring methods.

Apprentice electricians spend approximately 8,000 hours in training. They are paid about half the salary of journeymen and receive pay increases every six months (journeymen are first-class electricians who have completed apprenticeship training or have gained sufficient experience in the field).

Organization and Structure

The construction industry is made up of numerous entities. Its purpose is to provide construction services to the public and private sectors using highly advanced equipment, materials, and techniques.

Company Policies and Procedures

The key to any construction organization is its company policies and procedures, which provide the means for carrying out management processes, aid in decision making and problem solving, and provide consistency to business operation.

Policies and procedures vary widely from one contractor to the next; but most fall under the following general headings:

- Payroll practices
- Safety
- Drug and alcohol use
- Personnel
- Work schedule
- Benefits

Company procedures are designed to improve overall efficiency, to reduce the cost of running a business, to standardize the ways in which work is performed, and to create a smooth work atmosphere.

Some smaller contractors may give information about company procedures orally, but most contractors provide this information in writing. Sources of this information inclued the following:

1. ***Company Manual:*** This publication provides general information about the company. It may describe the history and function of the company and its organizational structure. It may also give the names of company and departmental or division executives.

2. ***Employee Manual:*** This publication may be separate from the company manual or included in it. In some cases, employee information may simply be posted on a bulletin board.

   This manual provides information or policies regarding work schedules, vacations, sick leave, holidays, company benefits, and so on. Some firms provide a separate manual for each job or trade.

3. ***Office Manual:*** This publication describes how specific job duties are performed. It may also include information about procedures for typing correspondence, billing, paying bills, ordering materials and equipment, answering the telephone, receiving guests, and similar office activities.

4. ***Job Manual:*** This publication usually describes each job, the firm's expectations of employees, and the supervisor for each job category.

Chapter 2

CODES AND STANDARDS

A good knowledge of the *National Electrical Code®* *(NEC)* is one of the first requirements in becoming a *trained electrical worker*. In fact, the *NEC* is probably the most widely used and generally accepted code in the world. It is used as an electrical installation, safety, and reference guide in the United States, and in many other parts of the world as well.

The *NEC*, originally prepared in 1897, is frequently revised to meet changing conditions, improved equipment and materials, and new fire hazards. It is a result of the best efforts of electrical engineers, manufacturers of electrical equipment, insurance underwriters, firefighters, and other concerned experts throughout the country.

The *NEC* is now published by the National Fire Protection Association (NFPA), Batterymarch Park, Quincy, Massachusetts 02269. It contains specific rules and regulations intended to help in the practical safeguarding of persons and property from hazards arising from the use of electricity.

Although the *NEC* itself states, "This Code is not intended as a design specification nor an instruction manual for untrained persons," it does provide a sound basis for the study of electrical installation procedures — under the proper guidance.

The *NEC* has become the bible of the electrical construction industry, and anyone involved in electrical work, in any capacity, should obtain an up-to-date copy, keep it handy at all times, and refer to it frequently.

This chapter covers the key terms and basic layout of the *NEC*. A brief review of the individual *NEC* sections that apply to electrical systems will be covered. This chapter, however, is not a substitute for the *NEC*. You need a copy of the most recent edition and it should be kept handy at all times. The more you know about the *NEC*, the more you are likely to refer to it.

NEC Terminology

There are two basic types of rules in the *NEC: mandatory rules* and *advisory rules*. Here is how to recognize the two types of rules and how they relate to all types of electrical systems.

- Mandatory rules — All mandatory rules have the word shall in them. The word *shall* means must. If a rule is mandatory, you *must* comply with it.
- Advisory rules — All advisory rules have the word should in them. The word *should,* in this case, means recommended but not necessarily required.

Be alert to local amendments to the *NEC*. Local ordinances may amend the language of the *NEC*, changing it from *should* to *shall*. This means that you must do in that county or city what may only be recommended in some other area. The office that issues building permits will either sell you a copy of the code that's enforced in that county or tell you where the code can be obtained. In rare instances, the electrical inspector having jurisdiction over the area may issue these regulations verbally.

There are a few other "landmarks" that you will encounter while looking through the *NEC*. These are summarized in Figure 2-1, and a brief explanation of each follows:

Explanatory material: Explanatory material in the form of *Fine Print Notes* is designated (FPN). Where these appear, the FPNs normally apply to the *NEC* Section or paragraph immediately preceding the FPN.

Change bar: A change bar in the margins indicates that a change in the *NEC* has been made since the last edition. When becoming familiar with each new edition of the *NEC*, always review these changes. There are also several illustrated publications on the market that point out changes in the *NEC* with detailed explanations of each. Such publications make excellent reference material.

Mandatory rules are characterized by the use of the word:

SHALL

A recommendation or that which is advised but not required is characterized by the use of the word:

SHOULD

Explanatory material in the form of Fine Print Notes is designated:

(FPN)

A change bar in the margins indicates that a change in the NEC has been made since the last edition.

A bullet indictates that something has been deleted from the last edition of the NEC.

Figure 2-1: "Landmarks" used in the *NEC*.

Bullets: A filled-in circle called a "bullet" indicates that something has been deleted from the last edition of the *NEC*. Although not absolutely necessary, many electricians like to compare the previous *NEC* edition to the most recent one when these bullets are encountered, just to see what has been omitted from the latest edition. The most probable reasons for the deletions are errors in the previous edition, or obsolete items.

Extracted text: Material identified by the superscript letter "x" includes text extracted from other NFPA documents as identified in Appendix A of the *NEC*.

As you open the *NEC* book, you will notice several different types of text used. Here is an explanation of each.

1. ***Black Letters:*** Basic definitions and explanations of the *NEC*.

2. ***Bold Black Letters:*** Headings for each *NEC* application.

3. ***Exceptions:*** These explain the situations when a specific rule does not apply.

Exceptions are written in italics under the Section or paragraph to which they apply.

4. ***Tables:*** Tables are often included when there is more than one possible application of a requirement.

5. ***Diagrams:*** A few diagrams are scattered throughout the *NEC* to illustrate certain *NEC* applications.

NAVIGATING THE NEC

The *NEC* is divided into the Introduction (Article 90) and nine chapters. Chapters 1, 2, 3, and 4 apply generally; Chapters 5, 6, and 7 apply to special occupancies, special equipment, or other special conditions. These latter chapters supplement or modify the general rules. Chapters 1 through 4 apply except as amended by Chapters 5, 6, and 7 for the particular conditions.

Chapter 8 covers communications systems and is independent of the other chapters except where they are specifically referenced therein.

Chapter 9 consists of tables and examples.

There is also the *NEC* Contents at the beginning of the book and a comprehensive Index at the back of the book. You will find frequent use for both of these helpful "tools" when searching for various installation requirements.

Each chapter is divided into one or more Articles. Chapter 1 contains Articles 100 and 110. These Articles are subdivided into Sections. For example, Article 110 of Chapter 1 begins with Section 110-2. Some sections may contain only one sentence or a paragraph, while others may be further subdivided into lettered or numbered paragraphs such as (a), (1), (2), and so on.

NEC Article 90 contains the following sections:

- Purpose (90-1)
- Scope (90-2)
- Code Arrangement (90-3)
- Enforcement (90-4)
- Mandatory Rules and Explanatory Material (90-5)
- Formal Interpretations (90-6)
- Examination of Equipment for Safety (90-7)
- Wiring Planning (90-8)

Once you are familiar with Articles 90, 100, and 110 you can move on to the rest of the *NEC*. There are several key sections you will use often in servicing electrical systems. Let's discuss each of these important sections.

Wiring Design and Protection: Chapter 2 of the *NEC* discusses wiring design and protection, the information electrical technicians need most often. It covers the use and identification of grounded conductors, branch circuits, feeders, calculations, services, overcurrent protection and grounding. This is essential information for any type of electrical system, regardless of the type.

Chapter 2 is also a "how-to" chapter. It explains how to provide proper spacing for conductor supports, how to provide temporary wiring and how to size the proper grounding conductor or electrode. If you run into a problem related to the design/installation of a conventional electrical system, you can probably find a solution for it in this chapter.

Wiring Methods and Materials: Chapter 3 has the rules on wiring methods and materials. The materials and procedures to use on a particular system depend on the type of building construction, the type of occupancy, the location of the wiring in the building, the type of atmosphere in the building or in the area surrounding the building, mechanical factors and the relative costs of different wiring methods.

The provisions of this article apply to all wiring installations except remote control switching (Article 725), low-energy power circuits (Article 725), signal systems (Article 725), communication systems and conductors (Article 800) when these items form an integral part of equipment such as motors and motor controllers.

There are four basic wiring methods used in most modern electrical systems. Nearly all wiring methods are a variation of one or more of these four basic methods:

- Sheathed cables of two or more conductors, such as NM cable and BX armored cable (Articles 330 through 339)
- Raceway wiring systems, such as rigid and EMT conduit (Articles 342 to 358)
- Busways (Article 364)
- Cabletray (Article 318)

Article 310 in Chapter 3 gives a complete description of all types of electrical conductors. Electrical conductors come in a wide range of sizes and forms. Be sure to check the working drawings and specifications to see what sizes and types of conductors are required for a specific job. If conductor type and size are not specified, choose the most appropriate type and size meeting standard *NEC* requirements.

Articles 318 through 384 give rules for raceways, boxes, cabinets and raceway fittings. Outlet boxes vary in size and shape, depending on their use, the size of the raceway, the number of conductors entering the box, the type of building construction and atmospheric conditions of the areas. Chapter 3 should answer most questions on the selection and use of these items.

The *NEC* does not describe in detail all types and sizes of outlet boxes. But manufacturers of outlet boxes have excellent catalogs showing all of their products. Collect these catalogs. They are essential to your work.

Article 380 covers the switches, pushbuttons, pilot lamps, and similar devices that you will use to control electrical circuits or to connect portable equipment to electric circuits. Again, get the manufacturers' catalogs on these items. They will provide you with detailed descriptions of each of the wiring devices.

Article 384 covers switchboards and panelboards, including their location, installation methods, clearances, grounding and overcurrent protection.

Equipment for General Use

Chapter 4 of the *NEC* begins with the use and installation of flexible cords and cables, including the trade name, type letter, wire size, number of conductors, conductor insulation, outer covering and use of each. The chapter also includes fixture wires, again giving the trade name, type letter and other important details.

Article 410 on lighting fixtures is especially important. It gives installation procedures for fixtures in specific locations. For example, it covers fixtures near combustible material and fixtures in closets. The *NEC* does not describe how many fixtures will be needed in a given area to provide a certain amount of illumination.

Article 430 covers electric motors, including mounting the motor and making electrical connections to it. Motor controls and overload protection are also covered.

Articles 440 through 460 cover air conditioning and heating equipment, transformers and capacitors.

Article 480 gives most requirements related to battery-operated electrical systems. Storage batteries are seldom thought of as part of a conventional electrical system, but they often provide standby emergency lighting service. They may also supply power to security systems that are separate from the main ac electrical system.

Special Occupancies

Chapter 5 of the *NEC* covers special occupancy areas. These are areas where the sparks generated by electrical equipment may cause an explosion or fire. The hazard may be due to the atmosphere of the area or just the presence of a volatile material in the area.

Commercial garages, aircraft hangars and service stations are typical special occupancy locations.

Articles 500 through 501 cover the different types of special occupancy atmospheres where an explosion is possible. The atmospheric groups were established to make it easy to test and approve equipment for various types of uses.

Articles 501-4, 502-4 and 503-3 cover the installation of explosionproof wiring. An explosionproof system is designed to prevent the ignition of a surrounding explosive atmosphere when arcing occurs within the electrical system.

There are three main classes of special occupancy locations:

- Class I (Article 501): Areas containing flammable gases or vapors in the air. Class I areas include paint spray booths, dyeing plants where hazardous liquids are used and gas generator rooms.
- Class II (Article 502): Areas where combustible dust is present, such as grain-handling and storage plants, dust and stock collector areas and sugar-pulverizing plants. These are areas where, under normal operating conditions, there may be enough

combustible dust in the air to produce explosive or ignitable mixtures.

- Class III (Article 503): Areas that are hazardous because of the presence of easily ignitable fibers or flyings in the air, although not in large enough quantity to produce ignitable mixtures. Class III locations include cotton mills, rayon mills and clothing manufacturing plants.

Article 511 and 514 regulate garages and similar locations where volatile or flammable liquids are used. While these areas are not always considered critically hazardous locations, there may be enough danger to require special precautions in the electrical installation. In these areas, the *NEC* requires that volatile gases be confined to an area not more than 4 ft above the floor. So in most cases, conventional raceway systems are permitted above this level. If the area is judged critically hazardous, explosionproof wiring (including seal-offs) may be required.

Article 520 regulates theaters and similar occupancies where fire and panic can cause hazards to life and property. Drive-in theaters do not present the same hazards as enclosed auditoriums. But the projection rooms and adjacent areas must be properly ventilated and wired for the protection of operating personnel and others using the area.

Chapter 5 also covers residential storage garages, aircraft hangars, service stations, bulk storage plants, health care facilities, mobile homes and parks, and recreation vehicles and parks.

Special Equipment

The items in Chapter 6 are frequently encountered by commercial and industrial electrical workers.

Article 600 covers electric signs and outline lighting. Article 610 applies to cranes and hoists. Article 620 covers the majority of the electrical work involved in the installation and operation of elevators, dumbwaiters, escalators and moving walks. The manufacturer is responsible for most of this work. The electrician usually just furnishes a feeder terminating in a disconnect in a machine room which is usually in the bottom of the elevator shaft or else in the penthouse. The electrician may also be responsible for a lighting circuit to a junction box midway in the elevator shaft for connecting the elevator cage lighting cable and exhaust fans. Articles in Chapter 6 of the *NEC* give most of the requirements for these installations.

Article 630 regulates electric welding equipment. It is normally treated as a piece of industrial power equipment requiring a special power outlet. But there are special conditions that apply to the circuits supplying welding equipment. These are outlined in detail in Chapter 6 of the *NEC*.

Article 640 covers wiring for sound-recording and similar equipment. This type of equipment normally requires low-voltage wiring. Special outlet boxes or cabinets are usually provided with the equipment. But some items may be mounted in or on standard outlet boxes. Some sound-recording electrical systems require direct current, supplied from rectifying equipment, batteries or motor generators. Low-voltage alternating current comes from relatively small transformers connected on the primary side to a 120-V circuit within the building.

Other items covered in Chapter 6 of the *NEC* include: X-ray equipment (Article 660), induction and dielectric heat-generating equipment (Article 665) and machine tools (Article 670).

If you ever have work that involves Chapter 6, study the chapter before work begins. That can save a lot of installation time. Here is another way to cut down on labor hours and prevent installation errors. Get a set of rough-in drawings of the equipment being installed. It is easy to install the wiring outlet box or to install the right box in the wrong place. Having a set of rough-in drawings can prevent those simple but costly errors.

Special Conditions

In most commercial buildings, the *NEC* and local ordinances require a means of lighting public rooms, halls, stairways and entrances. There must be enough light to allow the occupants to exit from the building if the general building lighting is interrupted. Exit doors must be clearly indicated by illuminated exit signs.

Chapter 7 of the *NEC* covers the installation of emergency lighting systems. These circuits should be arranged so that they can automatically transfer to an alternate source of current, usually storage batteries or gasoline-driven generators. As an alternative in some types of occupancies, you can connect them to the supply side of the main service so disconnecting the main service switch would not disconnect the emergency circuits. See Article 700. *NEC* Chapter 7 also covers a variety of other equipment, systems and conditions that are not easily categorized elsewhere in the *NEC*.

Chapter 8 is a special category for wiring associated with electronic communications systems including telephone and telegraph, radio and TV, fire and burglar alarms, and community antenna systems.

USING THE NEC

Once you become familiar with the *NEC* through repeated usage, you will generally know where to look for a particular topic. While this chapter provides you with an initial familiarization of the *NEC* layout, much additional usage experience will be needed for you to feel comfortable with the *NEC's* content. Until you have gained this practical experience, you can still use the *NEC* to answer questions pertaining to your work. Here's how to locate information on a specific subject.

Step 1. Look through the Contents. You may spot the topic in a heading or subheading. If not, look for a broader, more general subject heading under which the specific topic may appear. Also look for related or similar topics. The Contents will refer you to a specific page number.

Step 2. If you do not find what you're looking for in the Contents, go to the Index at the back of the book. This alphabetic listing is finely divided into different topics. You should locate the subject here. The Index, however, will refer to you either an Article or Section number (not a page number) where the topic is listed.

Step 3. If you cannot find the required subject in the Index, try to think of alternate names. For example, instead of wire, look under conductors; instead of outlet box, look under boxes, outlet, and so on.

The *NEC* is not an easy book to read and understand at first. In fact, seasoned electricians sometimes find it confusing. Basically, it is a reference book written in a legal, contract-type language and its content does assume prior knowledge of most subjects listed. Consequently, you will sometimes find the *NEC* frustrating to use because terms aren't always defined, or some unkown prerequisite knowledge is required. To minimize this problem, it is recommended that you obtain one of the several *NEC* supplemental guides that are designed to explain and supplement the *NEC*. One of the best is *McGraw-Hill's Illustrated Index to the 1996 NEC*, available directly from McGraw-Hill, Inc., 11 W. 19th St., New York, NY 10011, or from your local book store.

Practical Application

Let's assume that you are installing track lighting in a commercial office. The owner wants the track located behind the curtain of their sliding glass patio doors. To determine if this is an *NEC* violation or not, follow these steps:

Step 1. Turn to the Contents of the *NEC* book.

Step 2. Find the chapter that would contain information about the general application you are working on. For this example, Chapter 4 — Equipment for General Use — should cover track lighting.

Step 3. Now look for the article that fits the specific category you are working on. In this case, Article 410 covers lighting fixtures, lampholders, lamps, and receptacles.

Step 4. Next locate the *NEC* Section within the *NEC* Article 410 that deals with the specific application. For this example, refer to Part R — Lighting Track.

Step 5. Turn to the page listed.

Step 6. Read *NEC* Section 410-100, Definition to become familiar with track lighting.

Continue down the page with *NEC* Section 410-101 and read the information contained therein. Note that paragraph (c) under *NEC* Section 410-101 states the following:

"(c) Locations Not Permitted. Lighting track shall not be installed (1) where subject to physical damage; (2) in wet or damp locations; (3) where subject to corrosive vapors; (4) in storage battery rooms; (5) in hazardous (classified) locations; (6) where concealed; (7) where extended through walls or partitions; (8) less than 5 feet above the finished floor except where protected from physical damage or track operating at less than 30 volts RMS open-circuit voltage."

Step 7. Read *NEC* Section 410-101, paragraph (c) carefully. Do you see any conditions that would violate any *NEC* requirements if the track lighting is installed in the area specified? In checking these items, you will probably note condition (6), "where concealed." Since the track lighting is to be installed behind a curtain, this sounds like an *NEC* violation. But let's check further.

Step 8. Let's get an interpretation of the *NEC's* definition of "concealed." Therefore, turn to Article 100 — Definitions and find the main term "concealed." It reads as follows:

"Concealed: Rendered inaccessible by the structure or finish of the building...."

Step 9. After reading the *NEC's* definition of "concealed," although the track lighting may be out of sight (if the curtain is drawn), it will still be readily accessible for maintenance. Consequently, the track lighting is really not concealed according to the *NEC* definition.

When using the *NEC* to determine correct electrical-installation requirements, please keep in mind that you will nearly always have to refer to more than one Section. Sometimes the *NEC* itself refers the reader to other Articles and Sections. In some cases, the user will have to be familiar enough with the *NEC* to know what other *NEC* Sections pertain to the installation at hand. It's a confusing situation to say the least, but time and experience in using the *NEC* frequently will make using it much easier.

Now let's take another example to further acquaint you with navigating the *NEC*.

Suppose you are installing Type SE (service-entrance) cable on the side of a home. You know that this cable must be secured, but you aren't sure of the spacing between cable clamps. To find out this information, use the following procedure:

Step 1: Look in the *NEC* Table of Contents and follow down the list until you find an appropriate category.

Step 2: Article 230 under Chapter 3 will probably catch your eye first, so turn to the page where Article 230 begins in the *NEC*.

Step 3: Glance down the section numbers, 230-1, Scope, 230-2, Number of Services, etc., until you come to Section 230-51, Mounting Supports. Upon reading this ection, you will find in paragraph (a) — Service-Entrance Cables — that "Service-entrance cable shall be supported by straps or other approved

> means within 12 inches (305 mm) of every service head, gooseneck, or connection to a raceway or enclosure and at intervals not exceeding 30 inches (762 mm)."

After reading this section, you will know that a cable strap is required within 12 in of the service head and within 12 in of the meter base. Furthermore, the cable must be secured in between these two termination points at intervals not exceeding 30 in.

DEFINITIONS

Many definitions of terms dealing with the *NEC* may be found in *NEC* Article 100. However, other definitions are scattered throughout the *NEC* under their appropriate category. For example the term *lighting track*, as discussed previously, is not listed in Article 100. The term is listed under *NEC* Section 410-100 and reads as follows:

> *"Lighting track is a manufactured assembly designed to support and energize lighting fixtures that are capable of being readily repositioned on the track. Its length may be altered by the addition or subtraction of sections of track."*

Regardless of where the definition may be located (in Article 100 or under the appropriate *NEC* Section elsewhere in the book) the best way to learn and remember these definitions is to form a mental picture of each item or device as you read the definition. For example, turn to the first page of Article 100 — Definitions, scan down the page until you come to the term "Appliance." Read the definition and then try to form a mental picture of what appliances look like. Some of the more common appliances appear in Figure 2-2. They should be familiar to everyone.

Continue scanning down the page until you come to the term "Attachment Plug (Plug Cap) (Cap)." After reading the definition, you will probably have already formed a mental picture of attachment plugs.

Each and every term listed in the *NEC* should be understood. Know what the item looks like and how it is used on the job. If a term is unfamiliar, try other reference books such as manufacturers' catalogs for an illustration of the item. Then research the item further to determine its purpose in electrical systems. Once you are familiar with all the common terms and definitions found in the *NEC*, navigating through the *NEC* (and understanding what you read) will be much easier.

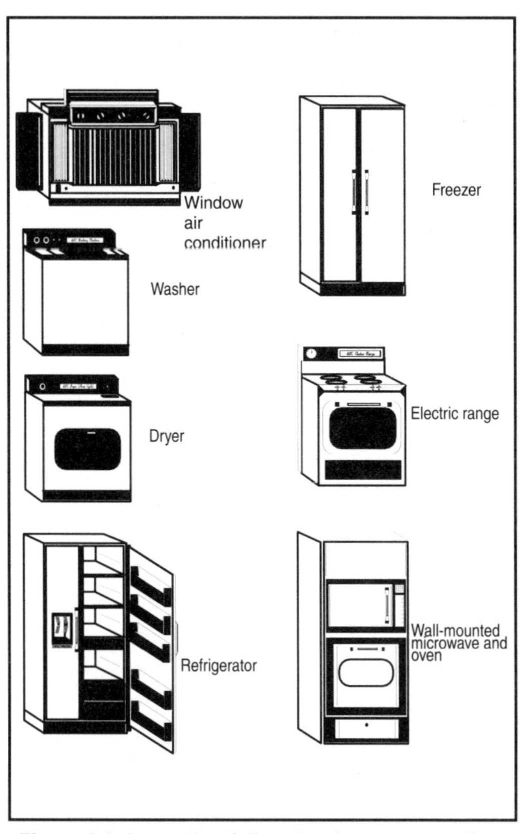

Figure 2-2: Items that fall under the category of "appliances" in *NEC* Article 100 — Definitions.

TESTING LABORATORIES

Many definitions may be found in Article 100. You should become familiar with the definitions. Since a copy of the lastest *NEC* is compulsory for any type of electrical wiring, there is no need to duplicate them here. However, here are two definitions that you should become especially familiar with:

- ***Labeled*** - Equipment or materials to which has been attached a label, symbol or other identifying mark of an organization acceptable to the authority having jurisdiction and concerned with product evaluation, that maintains periodic inspection of production of labeled equipment or materials, and by whose labeling the manufacturer indicates compliance with appropriate standards or performance in a specified manner.
- ***Listed*** - Equipment or materials included in a list published by an organization acceptable to the authority having jurisdiction and concerned with product evaluation, that maintains periodic inspection of production of listed equipment or materials, and whose listing states either that the equipment or material meets appropriate designated standards or has been tested and found suitable for use in a specified manner. Besides installation rules, you will also have to be concerned with the type and quality of ma-

terials that are used in electrical wiring systems. Nationally recognized testing laboratories (Underwriters' Laboratories, Inc. is one) are product safety certification laboratories. They establish and operate product safety certification programs to make sure that items produced under the service are safeguarded against reasonable foreseeable risks. Some of these organizations maintain a worldwide network of field representatives who make unannounced visits to manufacturing facilities to countercheck products bearing their "seal of listings." See Figure 2-3.

However, proper selection, overall functional performance and reliability of a product are factors that are not within the basic scope of UL activities.

To fully understand the *NEC*, it is important to understand the organizations which govern it.

Nationally Recognized Testing Laboratory (NRTL)

Nationally Recognized Testing Laboratories are product safety certification laboratories. They establish and operate product safety certification programs to make sure that items produced under the service are safeguarded against reasonable foreseeable risks. NRTL maintains a worldwide network of field repre-

Figure 2-3: Underwriters' Laboratories listing mark.

sentatives who make unannounced visits to factories to countercheck products bearing the safety mark.

National Electrical Manufacturers Association (NEMA)

The National Electrical Manufacturers Association was founded in 1926. It is made up of companies that manufacture equipment used for generation, transmission, distribution, control, and utilization of electric power. The objectives of NEMA are to maintain and improve the quality and reliability of products; to ensure safety standards in the manufacture and use of products; to develop product standards covering such matters as naming, ratings, performance, testing, and dimensions. NEMA participates in developing the *NEC* and the National Electrical Safety Code and advocates their acceptance by state and local authorities.

National Fire Protection Association (NFPA)

The NFPA was founded in 1896. Its membership is drawn from the fire service, business and industry, health care, educational and other institutions, and individuals in the fields of insurance, government, architecture, and engineering. The duties of the NFPA include:

- Developing, publishing, and distributing standards prepared by approximately 175 technical committees. These standards are intended to minimize the possibility and effects of fire and explosion.
- Conducting fire safety education programs for the general public.
- Providing information on fire protection, prevention, and suppression.
- Compiling annual statistics on causes and occupancies of fires, large-loss fires (over

1 million dollars), fire deaths, and firefighter casualties.

- Providing field service by specialists on electricity, flammable liquids and gases, and marine fire problems.
- Conducting research projects that apply statistical methods and operations research to develop computer modes and data management systems.

The Role of Testing Laboratories

Testing laboratories are an integral part of the development of the code. The NFPA, NEMA, and NRTL all provide testing laboratories to conduct research into electrical equipment and its safety. These laboratories perform extensive testing of new products to make sure they are built to code standards for electrical and fire safety. These organizations receive statistics and reports from agencies all over the United States concerning electrical shocks and fires and their causes. Upon seeing trends developing concerning association of certain equipment and dangerous situations or circumstances, this equipment will be specifically targeted for research.

Summary

The *National Electrical Code* specifies the minimum provisions necessary for protecting people and property from hazards arising from the use of electricity and electrical equipment. Anyone involved in any phase of the electrical industry must be aware of how to use and apply the code on the job. Using the *NEC* will help you to safely install and maintain the electrical equipment and systems that you come into contact with.

References

For a more advanced study of topics covered in this chapter, the following works are suggested:

National Electrical Code 1996 Handbook, NFPA, Quincy, MA.

Illustrated Guide to the 1996 NEC, Craftsman Book Company, 1996, Carlsbad, CA.

Illustrated Changes of the 1996 NEC, ATP, Homewood, IL.

Illustrated Pocket Guide to the 1996 NEC Tables, McGraw-Hill, Inc., New York, NY.

Illustrated Index to the 1996 NEC, McGraw-Hill, Inc., New York, NY.

McGraw-Hill's National Electrical Code Handbook, McGraw-Hill, Inc., New York, NY.

Chapter 3

TOOLS OF THE TRADE

Hand tools, except for specialized ones, are the responsibility of the electrician. Electrical contractors are normally responsible for furnishing all power and mechanical tools required for the installation of any electrical system. However, it is the electrician's responsibility to learn how to use all relative tools, and also how to care for them before, during, and after use.

This care covers the basic tools required for the majority of electrical installation, along with hints on their use.

All tools must also be in good repair. Tools meant to cut, for example, should be kept sharp. Dull tools can cause more accidents than sharp ones. Sharp tools are more efficient and allow workers to do their jobs better.

Hand Tools

The hand tools commonly used by electricians as set forth in the majority of labor agreements are as follows: (1 of each except for Channel locks or pump pliers where 2 pairs are normally required):

Tool box
Channel locks
6" Screwdriver
6' Folding rule
2" Conduit reamer
Expansion bit
Combination square
½" Cold chisel
½" Wood chisel
10" Crescent wrench
Keyhole saw
Gripping screwdriver
Small architect's scale
8" Level
Long nose pliers
Hacksaw frame
Fuse puller
8" side-cutting plier
10" Screwdriver
Claw hammer
Voltage tester
Brace
Bit extension
Center punch
Tap wrench
6" Crescent wrench
50' Steel tape
Phillip's screwdriver
10" Tin snips
Electrician's knife
Diagonal pliers
Lock
10" Mill file
Flashlight (preferably wired for testing continuity)

This list of hand tools is the minimum essential hand tools needed in order to perform good work on the usual kind of electrical installations. Any other tools needed are normally supplied by the contractor.

Knowing the Tools

As with other types of electrical materials, it is also necessary for all personnel to have a good knowledge of the various kinds of tools and installation equipment that are necessary to perform various electrical installations. The following illustrations depict most hand tools used by electricians.

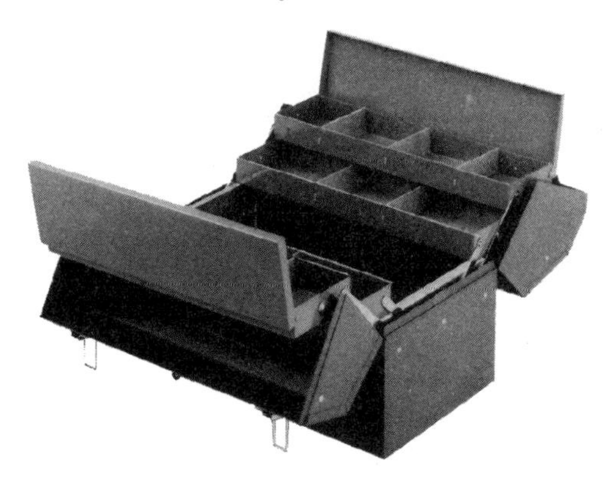

Figure 3-1: Klein 4-tray tool box.

While workers with any amount of experience at all are usually adept at using the installation tools and equipment of the trade, there may be instances where workers may have acquired bad habits or awkward movements, which, if corrected, increase their pro-

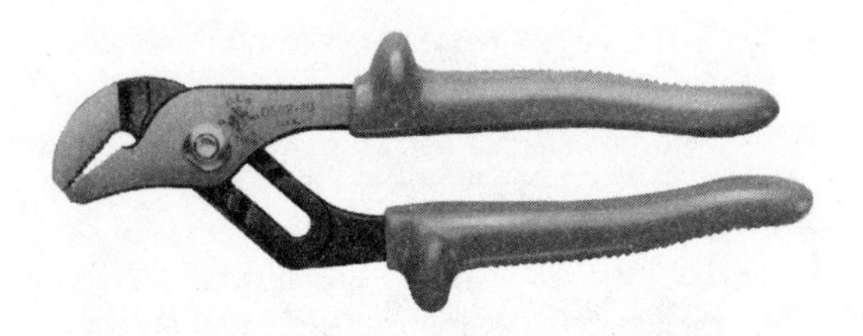

Figure 3-2: Klein insulated pump pliers, often called "Channel-locks" on the job.

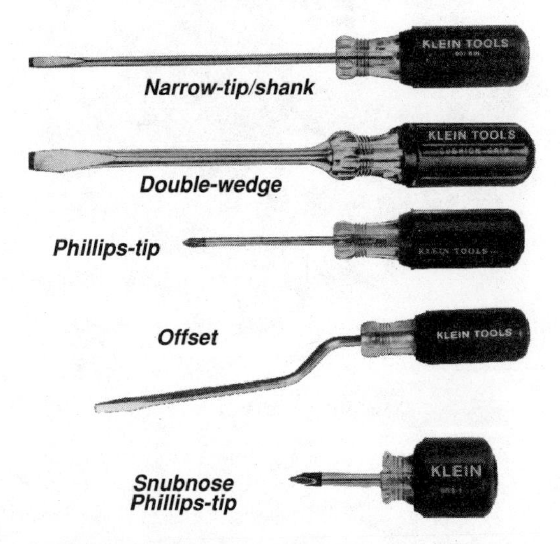

Figure 3-3: Several types and sizes of screwdrivers are needed by every electrician.

ductive efficiency. For example, bad habits common to many workers include:

1. Not cleaning out threading dies or keeping them tight in the stock.

2. Continuing to use dull dies and hack saw blades.

3. Not using sufficient cutting fluid while cutting threads.

4. Throwing diestocks and hack saws to the deck or floor instead of leaning or hanging them on the vise stand.

5. Assuming an awkward stance when turning a diestock.

6. Putting too much pressure on a hack saw.

Figure 3-4: Six-ft folding rule.

Figure 3-5: Many electricians prefer the 10-ft retractable metal tape to the 6-ft folding rule.

7. Pulling wire with a tangle of fish-steel about their feet.

8. Using a screwdriver with too small or too large a bit.

9. Failing to properly adjust crescent type wrenches or using ill-fitting end wrenches with resultant damage to boltheads and injured knuckles when the wrench slips.

10. Failing to properly adjust pipe wrenches or tongs resulting in "chewed-up" conduit and possible injury to the worker if the wrench or tong slips under a heavy pull.

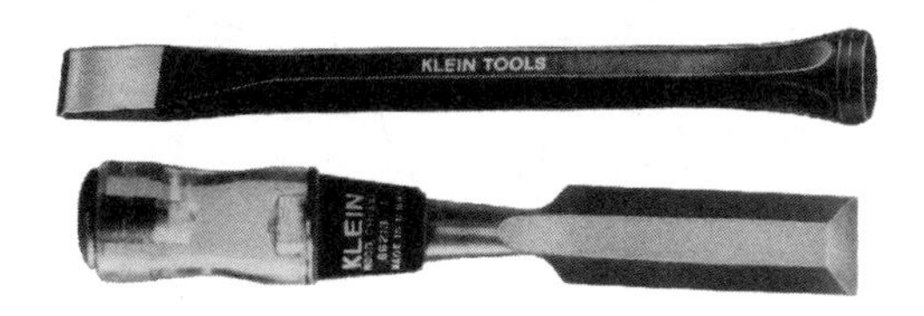

Figure 3-6: A cold chisel (top) and a wood chisel (bottom) are usually included in an electrician's tool box.

11. Using tools for the wrong purpose, for example, using a pair of pliers and a screw driver as a hammer and punch.

Anyone involved with electrical construction should be alert to such bad habits and awkward movements.

Figure 3-7: While a keyhole saw will find occasional use, if much cutting is to be done, the contractor normally furnishes electricians with either a sabre saw or else a Milwaukee Sawzall.

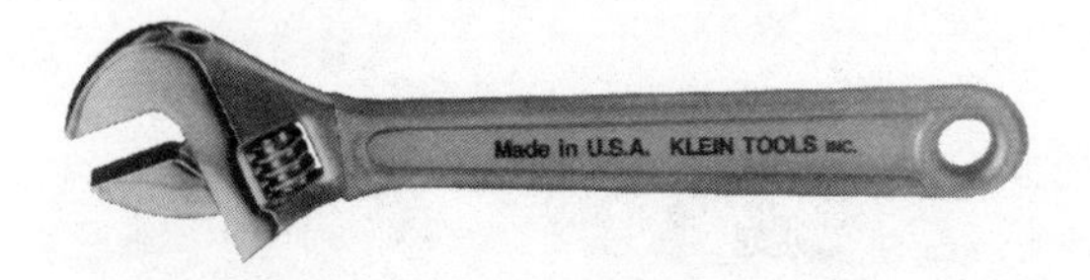

Figure 3-8: Typical adjustable-end wrench, commonly called "Crescent wrench."

Figure 3-9: Klein aluminum torpedo level.

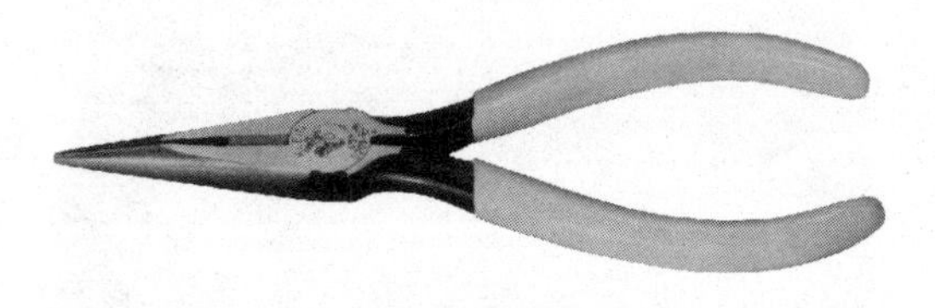

Figure 3-10: Klein long-nose pliers.

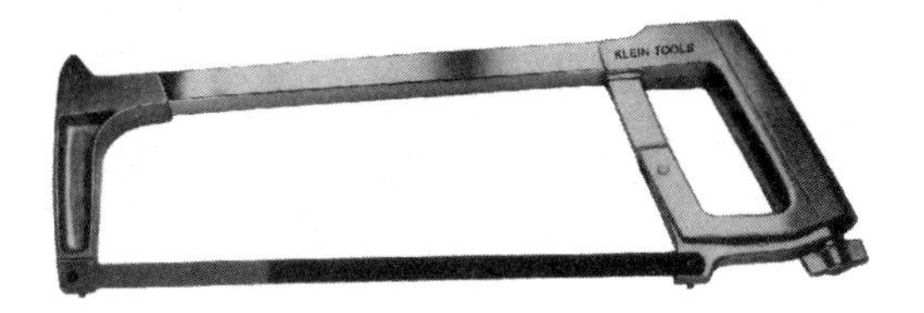

Figure 3-11: Klein hacksaw frame.

Figure 3-12: Klein side-cutting pliers.

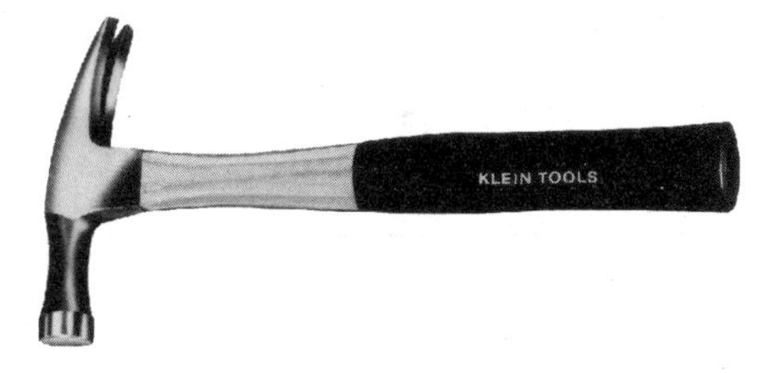

Figure 3-13: Typical electrician's claw hammer.

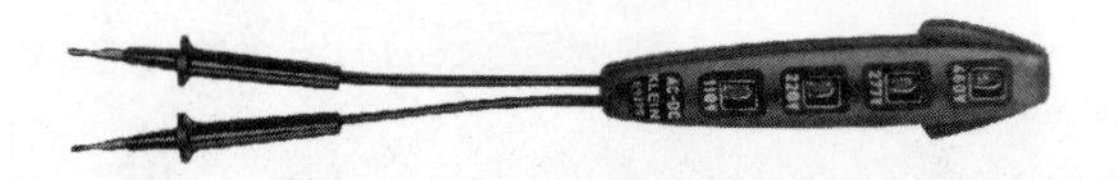

Figure 3-14: Klein multivolt twin-lead tester.

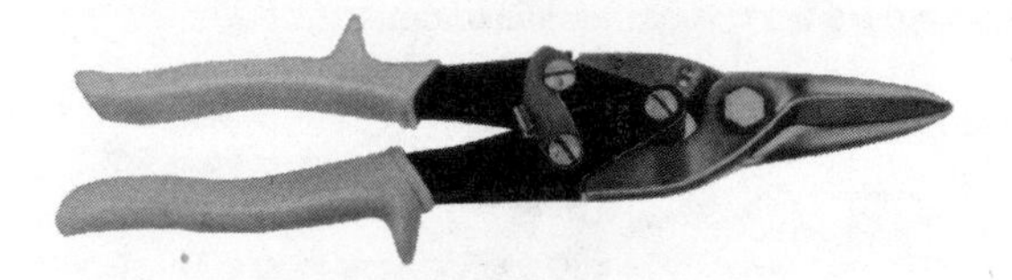

Figure 3-15: Typical tin snips.

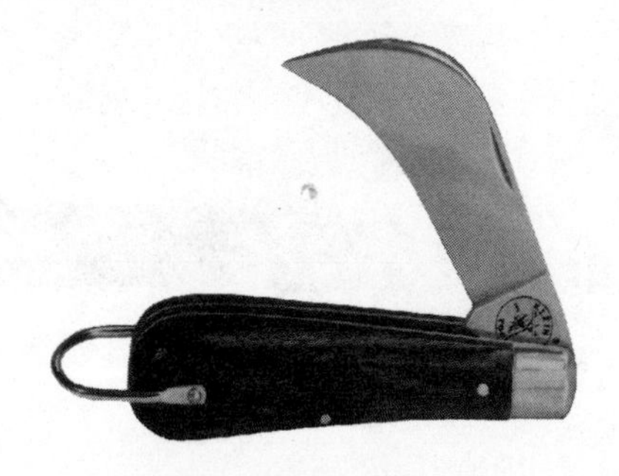

Figure 3-16: Klein slitting pocket knife.

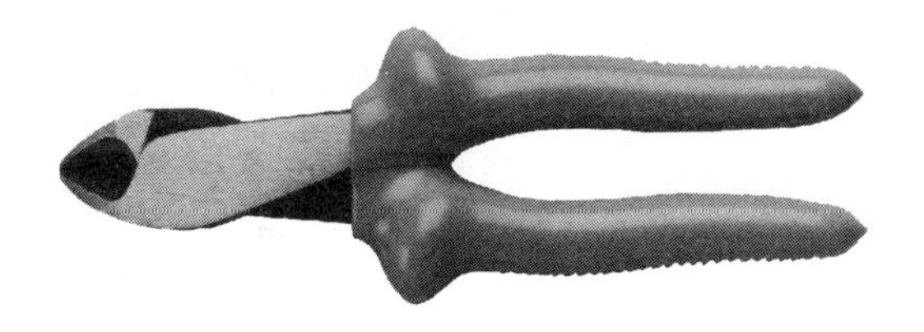

Figure 3-17: Klein insulated diagonal-cutting pliers.

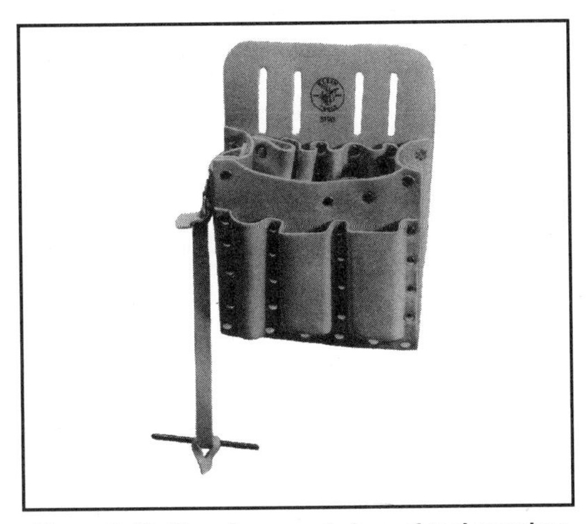

Figure 3-18: Many types and sizes of tool pouches are available; choose one that suits your needs the best.

Many new and improved hand tools and power operated tools are being introduced into the industry, which can be used to increase productivity, lessen fatigue and reduce injuries. Many otherwise experienced workers or contractors may not be familiar with the application and uses of some of these items. For a complete listing of electrical tools and materials, obtain a copy of Gene Whitson's *Handbook of Electrical Construction Tools and Materials*, available from McGraw-Hill, Inc. , New York, NY.

Proper Care and Use of Tools

The following is a brief summary of tool use and maintenance; other techniques will follow:

1. Sufficient space should be provided for, to protect all tools, equipment, and clothing.

2. Company tools should be used only for the purpose for which they are intended.

3. Chisels, star drills, drill, and all cutting tools are to be kept sharp. All impact tools should be kept dressed to prevent "mushrooming. "

4. The right size and proper type wrenches must be used.

5. No power tools should be hung by hose or cord.

6. All work should be conducted in a safe manner and all equipment should be so maintained, handled, and stored so as to avoid any danger to persons employed on the job.

7. Safety equipment should be handled and stored in the proper places and equipment such as rubber gloves and blankets should be electrically tested and tagged at least every two months and should be air tested before each use.

8. Don't throw tools or material from scaffolds, platforms, or walkways. Use hand lines to lower such items.

In any electrical business, certain personnel are usually designated to be responsible for the maintenance and replacement of the tools and equipment. This procedure ensures that tools will always be in satisfactory condition for use on the job to avoid lost time of workers in attempting to perform their work with badly worn or broken tools or equipment that are not in proper operating condition.

Electrical contractors cannot expect workers to perform satisfactory work in a normal length of time when they are provided with tools and equipment in an unsatisfactory or inoperative condition. The psychological reaction is also bad, as even first-class workers will take the attitude that if the contractor does not take the pains to provide them with satisfactory tools, why should they be concerned about performing the work in a workmanlike manner and in the minimum amount of time.

Replacement of Tools and Equipment

When the condition of tools and equipment becomes such that they cannot be restored to satisfactory condition by maintenance or repairing, or they become obsolete, they should be replaced immediately. Retention of such items in the tool and equipment stock gives a false impression as to what items are on hand, and there is always a possibility that they will be shipped to a job for one last period of use. This is really false economy if one considers the possibility of lost time of workers attempting to use them or possible injury or loss of life.

Many contractors will make an effort to obtain the most modern and adaptable tools and equipment which will "stand up" the longest and will require the least maintenance and repair. They know that the purchase of an inferior item because its initial cost may be slightly lower than a better item is false economy.

General Classification of Tools

1. Shop tools and equipment
2. Hand tools — small and medium sizes
3. Expendable tools
4. Conduit and other raceway tools
5. Portable power tools
6. Powder actuated tools
7. Wire and cable installation tools
8. Underground installation tools
9. Cable splicing tools
10. Wearing apparel

11. Mechanical trenching and digging equipment

12. Line construction tools and equipment

13. Trucks and other automotive equipment

14. Safety equipment

15. Testing and measuring equipment

16. Special facilities

Certain items of tools and equipment are more generally confined to, and used in the shop to process, fabricate, or test electrical materials, fixtures, operating equipment and the like. In most cases, most such work may be performed before the materials are delivered to the job site, or in some instances, material may be returned from the job to the shop, the necessary work performed, and the item then returned to the job.

When the size or type of the job is large enough, a job-site shop may be set up equipped, to at least some extent, with shop-type tools.

The size of the contractor's operations may be a determining factor as to the type of tools and equipment kept in the shop rather than being placed on the jobs. For example, a relatively small operation might maintain a power hack saw and, say a pipe threading machine in the shop only, whereas a larger contractor would not only maintain such equipment in the shop but find it advisable to also place such items of equipment on certain large jobs.

While it is not possible to draw an absolute line of distinction between items of tools and equipment that are maintained in a shop or used on jobs, the following list is representative of those maintained in a well-equipped shop.

- Work benches
- Drill press
- Electric tool grinder
- Power hack saw
- Power threader
- Hydraulic bender
- Lathe
- Heavy metal shear and punch
- Electric welder and equipment
- Forge
- Anvil
- Air compressor
- Machinist vise
- Pipe vise
- Set of taps

- Set of socket wrenches
- Set of rod stock and dies
- Hand truck
- Dollies
- Power threading tools
- Hand and power bending tools
- Conduit cutting devices
- Fish tapes and vacuum fish
- Set of steel lettering tape machines, and numbering dies
- Oxyacetyline tank, gauges
- Wire and cable reeling and burning equipment, and measuring equipment

The use of certain items of tools and equipment is not confined to the performance of any particular phase of electrical construction. At least some such items are usually required on any job. These items include the following:

- Padlocks
- Step ladders
- Extension ladders
- Rolling scaffolds
- Chain hoists
- Wagon trucks
- Dollies

- Blocks and falls
- Tarpaulins
- Transit and level
- Walkie-talkie
- Extension or drop cords
- Tool and storage boxes and/or tool shed
- Ladders
- Equipment securing chains

Specific Types of Tools

Conduit and Raceway Tools

A large portion of the work on many electrical construction projects involve the installation of various raceways and conduit systems. The tools required for this work include the following:

- Pipe vise and stand
- Stocks and dies
- Conduit benders and hickeys
- Conduit cutters and reamers
- Mechanical benders
- Hydraulic benders

Electric Drills

The drill motor is the backbone of the portable electric tool industry. It takes on many forms, from drills to hammer-drills to diamond drills to magnetic drills, but the pistol drill or portable hand drill is perhaps the most common.

As you review the various types of drill motors, you will note that they run a large range of revolutions per minute (rpm), and also carry a range of such capacities from ¼ in through ½ in. What is not commonly understood is how these ratings are determined. Contrary to popular belief, the motor itself has nothing to do with the capacity rating of a drill motor. There is a table of standards that informs us that, say, a ¼-in high-speed drill bit can turn at 2000-2500 rpm under continuous operation in steel plate and will not damage the bit; that is, the bit will not become damaged or turn blue from excessive speed. The same applies to a ⅜-in high-speed bit turning at 650-1000 rpm. There is no reference to the motor since it is not determined whether the motor is capable of withstanding a continuous load of drilling in those capacities in steel. From this, it is easy to understand why an inexpensive drill can obtain the same capacity rating of a quality industrial rated drill.

Since all quality tools are designed for continuous operation within their rated capacity, you can quickly determine that a $\frac{1}{4}$-in drill motor should use a maximum high-speed bit of $\frac{1}{4}$ in for steel, or a $\frac{3}{8}$-in bit in a $\frac{3}{8}$-in drill motor. This same rule of thumb applies to all tools throughout the line. Capacity in hardwood is usually twice that in steel.

Vacuum Cleaners

Next to the common drill motor, probably nothing in the power tool line reaches a broader market than the vacuum. It is a very common item as one can be found in nearly every household in this country. In general, it is a high-speed motor that creates suction to lift and remove dirt, dust, and debris from floors, rugs, drapes, shelves, and machinery, as well as an endless list of other uses.

The applications for industrial vacuums challenge the imagination: for cleanup work, cleaning controls, removing lint, cleaning dust from drilled holes, for fish tape systems, and many other uses.

Two types of models are in common use: the dry pickup and the wet/dry models. Manufacturers' catalogs contain complete descriptions.

Electric Hammers

When man first attempted to penetrate concrete, it was necessary to obtain a star drill and heavy hammer. The star drill was hit once, turned slightly, and then hit again; this procedure was repeated until sufficient chips were removed to make a hole. Drilling a hole in this manner is hard work and as power tools were developed, a tool emerged that would do most of the work; namely, the electric hammer. It is still found in the industry for use in chipping, chiseling, scaling, cleaning of bricks, cleaning out motors, and so on.

When the common electric hammer was used, the operator had to turn the star drill as the hammer pounded. For even less effort on the operator's part, the rotary hammer was developed. These tools both hammered and rotated at high speeds, and were capable of drilling 1½ in with solid bits and 2½ in with core bits. Then the anchor manufacturers started producing anchors that would fill holes drilled by these rotary hammers. Exotic drop-in anchors required smaller holes, so there became a pressing need for smaller rotary hammers capable of drilling ¾-in diameter holes and smaller.

A hammer-drill is a simple dual-purpose tool, as its name implies. It is a high-speed hammer, and it is a drill. These small hammer-drills fell into the market-place for much the same reasons that medium-size hammers did. Anchor manufacturers developed a large range of drop-in anchors, which created a demand for a small high-speed, hammer-drill.

Portable Band Saw

There are many times when it is impractical to take work to a large band saw, so it is more practical to take the saw to the work. Workers in the electrical industry have found many uses for the portable band saw: cutting conduit, hanger rods, extrusions, bolts, cable, plastics and other materials.

Band saws use two rotating wheels with rubber tires that drive a continuous blade. Several different types of blades are available for cutting various materials, but the most common are blades made from carbon steel, alloy steel, and high-speed steel. Blades are available with 6, 8, 10, 14, 18, and 24 teeth per inch. In cutting metals, the rule of thumb is to always have three teeth in the material at all times. Using too coarse a blade can cause thin metals to hang up in the gullet between two teeth, and tear out a section of teeth.

Choose a blade with as few teeth per inch as possible to accomplish the cutting quickly, but keep in mind the rule of three teeth in the cut.

Miscellaneous Power Tools

The electrical industry, probably more so than other trades, has developed a power tool for practically all operations: pulling wire and cable, blowing or sucking fish tapes, drilling holes, bending conduit, cutting conduit, threading conduit, pushing conduit under roads, and many, many others. Any electrical contractor with sufficient work can benefit from any of them — provided they know the availability of the tools and the proper method of using them.

Chapter 4

OSHA AND ELECTRICAL SAFETY

Construction sites are hazardous places to work, but a thorough understanding of safe work practices and procedures will help to avoid injury and accidents.

Every worker should be concerned about safety — both on and off the job. Your failure to adopt recommended safety procedures can result in serious injury to yourself and your fellow workers, and can cause costly damage to equipment and property.

We often think that accidents happen to others, not to ourselves. But the construction industry, perhaps more than any other, requires constant dedication to safety. Be aware of all potential hazards and stay alert to them.

OSHA

The Federal Occupational Safety and Health Act, better known as OSHA, is a federal law enacted by the United States Congress to help insure safety in the workplace. The requirements of this law are administered by the Occupational Safety and Health Administration within the Department of Labor and are enforced either directly by OSHA or by the individual states, who may also have their own state codes with additional workplace safety standards above and beyond OSHA. The OSHA Safety and Health Standards for the Construction Industry, 29 CFR 1926/1910, is the Federal publication that enumerates the specific safety standards pertaining to construction. It is available through the Superintendent of Documents, U.S. Government Printing Office, Washington, D.C. 20402.

The basic intention of OSHA is to make employers aware of safety and health problems on the job and to make them responsible for creating a safe working environment for all employees. It encourages employers and employees to reduce workplace hazards and to implement new or improve existing safety and health programs.

Other goals of OSHA include:

- Providing research in occupational safety and health to develop innovative ways of dealing with occupational safety and health problems
- Establish "separate by dependent responsibilities and rights" for employers and employees for the achievement of better safety and health conditions

- Maintain a reporting and record-keeping system to monitor job-related injuries and illnesses
- Establish training programs to increase the number and competence of occupational safety and health personnel
- Develop mandatory job safety and health standards and enforce them effectively
- Provide for the development, analysis, evaluation and approval of state occupational safety and health programs

Electrical Contractors Covered by OSHA

OSHA applies to companies with 11 or more employees and extends to all employers and employees in the 50 states, the District of Columbia, Puerto Rico, and all other territories under Federal Government jurisdiction. If your employer is engaged in a business affecting commerce who has employees, but does not include the United States or any State or political subdivision of a State, he is subject to OSHA. OSHA applies to employers and employees in such varied fields as manufacturing, construction, longshoring, agriculture, law and medicine, charity and disaster relief, organized labor and private education. Such coverage includes religious groups to the extent that they employ workers for secular purposes.

Not covered under OSHA are:

- Self-employed persons
- Farms at which only immediate members of the farm employer's family are employed
- Working conditions regulated by other federal agencies under other federal statutes

The Electrical Worker and OSHA Safety Standards

In carrying out its duties, OSHA is responsible for broadcasting legally enforceable standards. OSHA standards may require conditions, or the adoption of use of one or more practices, means, methods or processes reasonably necessary and appropriate to protect workers on the job. It is the responsibility of employers to become familiar with standards applicable to their establishments and to ensure that employees have and use personal protective equipment when required for safety.

Employees must comply with all rules and regulations which are applicable to their own actions and conduct.

Where OSHA has not announced specific standards, employers are responsible for following the general duty clause. This clause states that each employer "Shall furnish . . . a place of employment which is free from recognized hazards that are causing or are likely to cause death or serious physical harm to his employees."

States with OSHA-approved occupational safety and health programs must set standards that are at least as effective as the federal ones.

Basic OSHA safety standards of 29 CFR 1926/1910, referred to earlier, not specifically electrical in context, but which apply to the electrical workers and their employers include:

- Safety apparel in the form of head, ear and eye protection
- Hardhats are required at areas where accidental head injury can be anticipated
- Tools must be safe to use under the anticipated construction environment
- Employers must take on the responsibility to ensure that proper medical treatment can be made available to the job site in a timely manner
- Employers must train their employees in being able to recognize potentially unsafe workplace conditions so that the employee can either correct the conditions himself or notify his employer of such conditions
- Toilet facilities must be provided as well as drinking water

Safe Motor Vehicle Standards

These standards are basically construction-site driving conditions. The employer must maintain the vehicle in a safe-driving condition and must provide a safe means of maneuvering the vehicle within the construction area.

Asbestos Standards

Asbestos can be encountered in most any construction project which is a remodeling project and/or construction of a new addition onto an existing building. OSHA has voluminous standards and regulations concerning asbestos. Exposure to asbestos in construction work is covered by 29 CFR1926.58. At all areas on a construction site where asbestos may present a health hazard, warning signs must be erected with the following information:

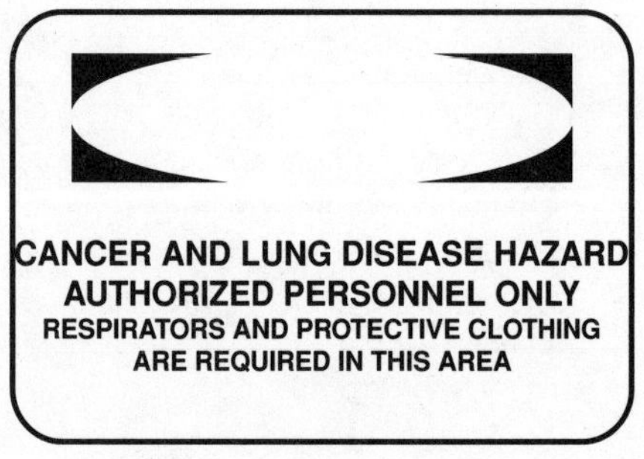

Inside the areas bounded by these signs are highly restrictive and specific regulations which must be observed by everyone. Do not enter such an area unless you have been trained in the handling of asbestos and your employer has deemed you qualified and knowledgeable in the handling of asbestos. Some of the OSHA training program requirements for the handling of asbestos include:

- Training prior to the employee's assignment to an asbestos area
- Education concerning the health effects of asbestos exposure including the relationship between smoking and asbestos exposure
- Education as to the quantity, location, manner of use, release, storage of asbestos, and the specific nature or operations which could result in exposure to asbestos
- Instruction as to the engineering controls and work practices associated with the employee's job assignment
- Instruction as to the specific procedure implemented to protect employees from exposure to asbestos, such as appropriate work practices, emergency and clean-up procedures, and personal protective equipment to be used

- Proper use of respirators and protective clothing
- The purpose and description of the medical surveillance program that is required

Workers should also be familiar with asbestos labels which are required to be affixed to all raw materials, mixtures, scrap, waste, debris, and other products containing asbestos fibers, or to their containers.

OSHA Recordkeeping

Employees should be aware of procedures that are required to be enacted by a company on behalf of the employee. Once an injury or a condition that potentially poses a health or safety threat arises, the employee should make a report to his or her employer. The employer is then required to investigate. Also, a record of injuries and/or job-related illnesses must be maintained and submitted to the local OSHA administrator. Employers of 11 or more employees must maintain records of occupational injuries and illnesses as they occur. Employers with fewer employees are exempt from keeping such records unless they are selected by the Bureau of Labor Statistics to participate in the Annual Survey of Occupational Injuries and Illnesses. The purposes of keeping records are to permit the survey material to be compiled, to help define high hazard industries, and to inform employees of the status of their employer's record.

Occupational Injury

Simple everyday worker mishaps that are only a temporary inconvenience or requiring only minor first aid are not going to credit an investigation or a disability check. Injuries that potentially will cause long-term effects and will require medical treatment are the ones that should be reported. A cut, fracture, sprain or amputation that results from a work-related accident or from exposure involving a single incident in the work environment constitutes an occupational injury.

Occupational Illness

An occupational illness is any abnormal condition or disorder, other than one resulting from an occupational injury, caused by exposure to environmental factors associated with employment. Included are acute and chronic illnesses or diseases which may be caused by inhalation, absorption, ingestion or direct contact with toxic substances or harmful agents.

Keeping Injury and Illness Records

All occupational illnesses must be recorded regardless of severity. All occupational injuries must be recorded if they result in:

- Death (must be recorded regardless of the length of time between the injury and death)
- One or more lost workdays
- Restriction of work or motion
- Loss of consciousness
- Transfer to another job
- Medical treatment (other than first aid)

Employers must keep injury and illness records for each establishment. An establishment is defined as a "single physical location where business is conducted or where services are performed."

OSHA On-Site Reminders

OSHA has recognized the fact that most workers (and employers) need to be constantly reminded of workplace safety. Some of the requirements of employers and penalties for non-compliance are printed on posters that are issued to contractors and must be prominently displayed in the workplace. The worker should read the poster carefully (including the fine print) so that he will be constantly reminded of his or her rights and the obligations to the workers by the employer. Posters may change from year to year; make certain that the poster is current.

Your Rights

As an employee, you have the right to:

- Review copies of appropriate OSHA standards, rules, regulations and requirements that the employer should have available at the workplace.
- Request information from your employer on safety and health hazards in the area, on precautions that may be taken, and on procedures to be followed if an employee is involved in an accident or is exposed to toxic substances.
- Receive adequate training and information on workplace safety and health hazards.
- Request the OSHA area director to investigate if you believe hazardous conditions or violations of standards exist in your workplace.
- Have your name withheld from your employer, upon request to OSHA, if you file a written and signed complaint.

- Be advised of OSHA actions regarding your complaint and have an informal review, if requested, of any decision not to inspect or to issue a citation.
- Have your authorized employee representative accompany the OSHA compliance officer during the inspection tour.
- Respond to questions from the OSHA compliance officer, particularly if there is no authorized employee representative accompanying the compliance officer.
- Observe any monitoring or measuring of hazardous materials and have the right to see these records, and your medical records, as specified under the Act.
- Have your authorized representative, or yourself, review the Log and Summary of Occupational Injuries (OSHA No. 200) at a reasonable time and in a reasonable manner.
- Request a closing discussion with the compliance officer following an inspection.
- Submit a written request to OSHA for information on whether any substance in your workplace has potentially toxic effects in the concentration being used and have your name withheld from your employer if you so request.

- Object to the abatement period set in the citation issued to your employer by writing to the OSHA area director within 15 working days of the issuance of the citation.
- Participate in hearings conducted by the Occupational Safety and Health Review Commission.
- Be notified by your employer if he or she applies for a variance from an OSHA standard, and testify at a variance hearing and appeal the final decision.
- Submit information or comment to OSHA on the issuance, modification, or revocation of OSHA standards and request a public hearing.

Signs

All workers should be aware of and understand all warning signs on the construction site; the instructions on these signs must be followed by everyone.

Danger Signs

A danger sign is signified by white lettering in a red oval on a black rectangular background as shown in Figure 4-1 on the next page. Danger signs indicate

Figure 4-1: Appearance of typical Danger Sign.

areas or machines that pose immediate hazards to workers and equipment. When this sign is encountered, the instruction must be followed exactly to avoid injury.

Caution Signs

Caution signs have a yellow background with black lettering. The word CAUTION is always printed at the top of the sign with the caution message below. For example, a caution sign stating "Open Door Slowly" could prevent a worker on the opposite side from becoming injured. *See* Figure 4-2.

Figure 4-2: Appearance of typical Caution Sign.

Safety First Signs

Safety First signs are similar to caution signs; that is, they offer warnings and suggestions that will help prevent accidents. This type of sign has an all-white background. SAFETY FIRST is superimposed in white letters on a green background, with the message below in black letters. *See* Figure 4-3.

Scaffolds

Scaffolds are commonly used on construction sites. They are meant to provide safe, secure, elevated work platforms for personnel and materials. Scaffolds should be designed and built to meet high safety standards. But normal wear and tear or overuse can

CLEAN UP
SPILLS

Figure 4-3: Appearance of typical Safety First Sign.

weaken a scaffold and make it unsafe. It is therefore important to inspect all parts of a scaffold before each use.

Three types of scaffolds are used in the construction industry:

- Manufactured
- Rolling
- Suspended

Manufactured scaffolds are made of painted steel, stainless steel, or aluminum, because these materials are stronger and more fire resistant than wood. They come in ready-made units that resemble the sections of a fence. The individual units are assembled on the job site.

A rolling scaffold is a manufactured scaffold with wheels on its legs so that it can be moved easily. The scaffold wheels are fitted with brakes to prevent movement while work is in progress.

A suspended scaffold is a platform supported by ropes or cables that are attached either to the top of some support structure or to beams extending from the side of a support structure. The platform is raised or lowered by pulling on the suspension ropes or cables using a hand crank or an electric motor.

Inspecting Scaffolds

A scaffold assembled for use should be tagged with either a green, yellow, or red tag.

A *green tag* identifies a scaffold that is safe for use and meets all OSHA standards.

A *yellow tag* identifies a scaffold that does NOT meet all standards. A yellow-tagged scaffold may be used, but workers must wear a safety harness and lanyard when they are on it. Other precautions may also apply.

A *red tag* identifies a scaffold that is being erected or taken down. Workers should never use a red-tagged scaffold.

Don't rely on the tags alone. Check the scaffold for bent, broken, or badly rusted tubes. Check for loose joints where the tubes are connected. Such problems present hazards that must be corrected before the scaffold can be used.

Make sure you know the scaffold's weight limit before using it. This weight should be compared with the total weight of the workers, tools, equipment, and material that will be put on the scaffold. Scaffold weight limits must NEVER be exceeded.

If a scaffold is higher than four feet, make sure that it is equipped with top rails, midrails, and toe boards. All connections must be pinned. Cross-bracing must be used. The working area must be completely planked. Cross-bracing does not eliminate the need for handrails.

If there is room under the scaffold for people to pass, the space between the toe board and top rail must be screened in to prevent tools and materials from falling off the work platform.

Manufactured or rolling scaffolds that are taller than four times the dimensions of their narrow base should be tied to the structure or guyed (wired) to the ground. For example, if a scaffold with a base of 4 ft by 6 ft is taller than 16 ft, it should be tied or guyed. *See* Figure 4-4 for safety requirements for scaffolds.

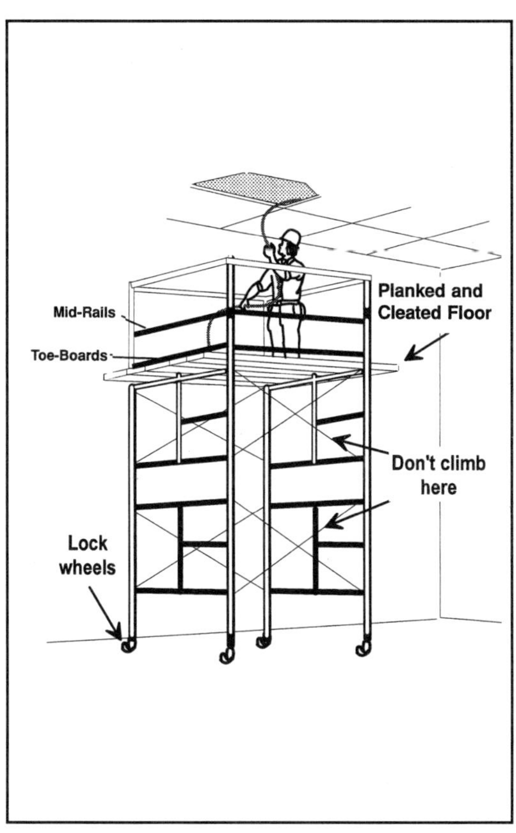

Figure 4-4: Some safety precautions to be used with scaffolds.

Electric Shock

Electricity is created when electrons move from a voltage source through a conductor that forms a complete path for the movement. This movement is called electric current. Silver, copper, steel, and aluminum are excellent conductors of electricity. Though less efficient, the human body is also a conductor.

Electric current flows back to its source along the easiest path. When the human body becomes the path, it receives an electric shock. If the shock is strong enough, it can be fatal.

Electric current is measured in amperes and milliamperes (a milliampere (mA) is one one-thousandth of an ampere).

The human body begins to feel electric shock when current flow approaches 1 mA. Between 1 and 5 mA, the body feels a tingling sensation. Above 5 mA, painful shock occurs, and at 50 mA or more, severe shock occurs that can cause death.

Body Resistance

The human body has relatively low resistance. For example, the resistance from one hand to the other or from one hand to a foot is about 300 to 600 ohms. However, when the body becomes a part of an electrical circuit, it can conduct a considerable amount of current. The amount of current depends on the condition of the skin and type of contact that the body makes with the voltage source.

For example, dry skin has an average contact resistance of 350,000 ohms. This relatively high resistance will minimize current flow. The amount of contact resistance, however, also depends on the amount of pressure that the skin makes with the voltage source. The greater the pressure and contact area, the lower the resistance and the greater the possibility of shock.

The moisture content of the skin also affects contact resistance. When the skin becomes wet from perspiration, water, or some other liquid, contact resistance drops from an average of 350,000 ohms to only 1,000 ohms. Again, the actual resistance will be even lower if the pressure and area of body contact increases, and thus the amount of current flow will be considerable. Obviously, working in wet conditions around electrical sources is more dangerous than working in dry conditions.

Very low-voltage sources, such as flashlight batteries or even a 12-V car or boat battery do not provide enough milliamperes to harm you. This voltage is simply too low to cause any significant amount of current to flow in the human body. However, even 40 V is enough to cause painful shock if the body is wet.

Most victims of electric shock are injured by the 120-V circuit — the most common voltage source. Construction workers sometimes become lax when working around 120-V lines, but use extreme caution when working with 480-V circuit, for example. In reality, both voltages can kill!

Preventing Electrical Shock

Use common sense and think of the safety as a personal attitude. It will help you to work a full career

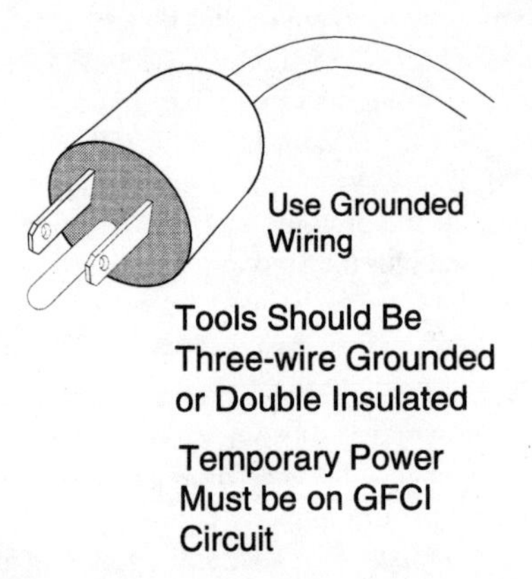

Figure 4-5: Make certain that power cords for portable hand tools have equipment grounding capabilities.

without a serious accident or injury. The precautions depicted in Figures 4-5 through 4-7 can further ensure a safe working career as an electrician.

Figure 4-6: The tag-and-lock method can be a life saver on certain construction projects.

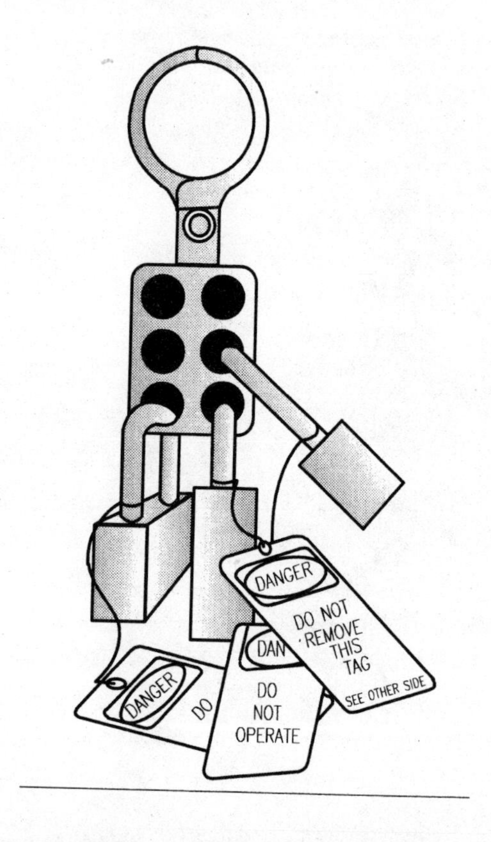

Figure 4-7: Equipment should be de-energized and locked-out before performing work on it.

Safety and Your Income

Life and health are priceless to everyone, but the simple fact is that if an injury causes you to miss work or become permanently disabled in some way, your income will be affected.

When you lose time, you lose income, but you continue to have expenses. Workers Compensation Insurance rarely pays enough to cover your losses. Your savings can be destroyed, and you can fall behind in your career. Lost time is not easily recovered.

Employability

If you want to change jobs, potential employers will often want to know about your safety record. Unsafe workers are costly to employers, as well as a danger to themselves and other workers.

The information in this chapter is not meant to discourage you from the electrical construction trade, but rather to help you understand and appreciate the need for safety precautions. These procedures will help keep you safe and protect the property you are working on throughout your career.

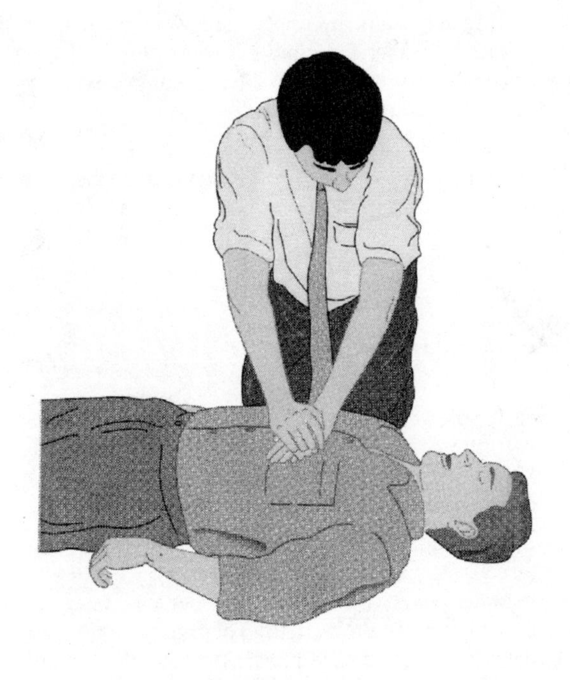

Figure 4-8: At least two members of each work team should be trained in CPR techniques.

Chapter 5

RACEWAY INSTALLATIONS

A raceway is any channel used for holding wires, cables, or busbars, which is designed and used solely for this purpose. Types of raceways include rigid metal conduit, intermediate metal conduit (IMC), rigid nonmetallic conduit, flexible metal conduit, liquid-tight flexible metal conduit, electrical metallic tubing (EMT), underfloor raceways, cellular metal floor raceways, cellular concrete floor raceways, surface metal raceways, wireways, and auxiliary gutters. Raceways are constructed of either metal or insulating material.

Raceways provide mechanical protection for the conductors that run in them and also prevent accidental damage to insulation and the conducting metal. They also protect conductors from the harmful chemical attack of corrosive atmospheres and prevent fire hazards to life and property by confining arcs and flame due to faults in the wiring system.

One of the most important functions of metal raceways is to provide a path for the flow of fault current to ground, thereby preventing voltage build-up on conductor and equipment enclosures. This feature, of course, helps to minimize shock hazards to personnel and damage to electrical equipment. To maintain this feature, it is extremely important that all metal raceway systems be securely bonded together into a continuous conductive path and properly connected to the electric service equipment-grounding system.

A box or fitting must be installed at:

- Each conductor splice point
- Each outlet, switch point, or junction point
- Each pull point for the connection of conduit and other raceways

Furthermore, boxes or other fittings are required when a change is made from conduit to open wiring. Electrical workers also install pull boxes in raceway systems to facilitate the pulling of conductors.

In each case — raceways, outlet boxes, pull and junction boxes — the *NEC* specifies specific maximum fill requirements; that is, the area of conductors in relation to the box, fitting, or raceway system.

Conduit Fill Requirements

The *NEC* provides rules for the maximum number of conductors permitted in raceways. In conduits, for either new work or rewiring of existing raceways, the maximum fill must not exceed 40 percent of the conduit cross-sectional area. In all such cases, fill is based on using the actual cross-sectional areas of the particular types of conductors used. Other derating rules specified by the *NEC* may be found in Article 310. For example, if more than three conductors are used in a single conduit, a reduction in current-carrying capacity is required. Ambient temperature is another consideration that may call for derating of wires below the values given in *NEC* tables.

The allowable number of conductors in a raceway system is calculated as percentage of fill as specified in Table 1 of *NEC* Chapter 9 (Figure 5-1). When using this table, remember that equipment grounding or bonding conductors, where installed, must be included when calculating conduit or tubing fill. The actual dimensions of the equipment grounding or bonding conductor (insulated or bare) must be used in the calculation.

Number of Conductors	1	2	Over 2
All Conductor Types	53%	31%	40%

Figure 5-1: Percent of cross section of conduit and tubing for conductors.

Conduit fill may be determined in one of two different ways:

- Calculating the fill as a percentage of the conduit's inside diameter (ID)
- Using tables in *NEC* Chapter 9

When determining the number of conductors (all the same size) for use in trade sizes of conduit or tubing ½ in through 6 in, refer to Tables C1 through C12 of *NEC* Appendix C. For example, let's assume that four 500 kcmil THHN conductors must be installed in a rigid conduit run. What size of conduit is required for four 500 kcmil THHN conductors?

Turn to *NEC* Appendix C, Table C8 and scan down the left-hand column until the insulation type (THHN) is found. Once the insulation type has been located, move to the second column (Conductor Size AWG/kcmil) and scan down this column until 500 kcmil is found. Now, scan across this row until the desired number of conductors (four in this case) is found. In this case, however, the number of conductors jumps from 3 to 5. Therefore, we will have to use the 5-conductor column. Move up this column to see the size of conduit required at the top of the page. In doing so, we can see that 3-in conduit is the size to use.

If compact conductors (all the same size) are used, refer to Table C8A in *NEC* Appendix C for trade sizes of conduit or tubing ½ through 6 inches. For example, let's assume that four 500 kcmil compact conductors are to be installed in a raceway system. What size of conduit is required?

Refer to Table C8A. Scan down the left-hand column until the insulation (THHN) is located. Once the insulation type has been located, move to the second column (Conductor Size AWG/kcmil) and scan down this column until 500 kcmil is found. Now, scan across this row until the desired number of conductors (four in this case) is found, or else the closest number to the desired number without being less (5-conductor column in this case). Now scan upward until the required conduit size is located. Again, the conduit size is 3 in — the same as in our previous example. Therefore, no savings in conduit size will be realized in this case if compact conductors are used. Furthermore, the compact conductors will be more costly and it will be to everyone's advantage to stick with conventional conductors in this situation.

When working with conductors larger than 750 kcmil or combinations of conductors of different sizes, Tables 4 through 8 of *NEC* Chapter 9 should be used to obtain the dimensions of conductors, conduit, and tubing. These tables give the nominal size of conductors or tubing for use in computing the required size of conduit or tubing for various combinations of conductors. The dimensions represent average conditions only, and variations will be found in dimensions of conductors and conduits of different manufacture.

Note: Where the calculated conductors, all of the same size (total cross-sectional area including insulation), include a decimal fraction, the next higher whole number must be used where the decimal is 0.8 or larger.

Let's take a situation where a 1200-A service-entrance is to be installed utilizing three parallel conduits, each containing three 500 kcmil THHN ungrounded conductors and one 350 kcmil grounded THHN conductor (neutral) as shown in Figure 5-2. What size of rigid conduit is required?

Step 1. Refer to *NEC* Chapter 9, Table 5 to determine the area (in in^2) of 500 kcmil THHN conductor. The area is found to be .7073 in^2.

Step 2. Since there are three conductors of this area in each conduit, .7073 is multiplied by 3 to obtain the total in^2 for all three conductors.

$$.7073 \times 3 = 2.1219$$

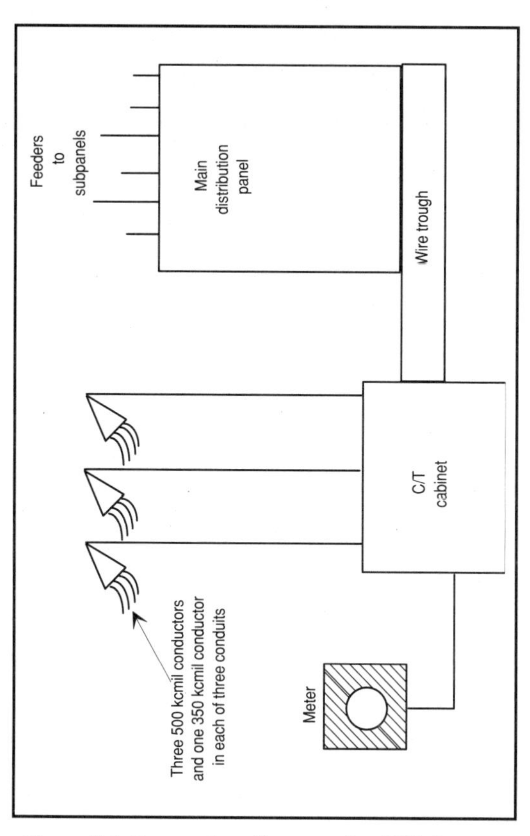

Figure 5-2: Power-riser diagram of a 1200 A service-entrance.

Step 3. Refer again to *NEC* Chapter 9, Table 5 to obtain the area of the one 350 kcmil grounded conductor in each conduit. The area is found to be .5242 in^2.

Step 4. Add the total area of the three 500 kcmil conductors (2.1219) plus the area of the one 350 kcmil conductor (.5242) to obtain the total area occupied by all four conductors in the conduit.

$$2.1219 + .5242 = 2.6461 \text{ in}^2$$

Step 5. Referring to Table 4 of *NEC* Chapter 9, we look in the "over 2 Cond. 40 percent" column and scan down until we come to a figure that is the closest to 2.6461 in^2 without going under this figure. In doing so, we find that the closest area is 3.00 in^2. Scan to the left from this figure (in the same column) and note that the conduit size is 3 in.

Therefore, each of the three conduits containing three 500 kcmil THHN conductors and one 350 kcmil THHN conductor must be 3 in ID to comply with the *NEC*.

Refer again to Figure 5-2 and note the conduit routed from the C/T cabinet to the meter location. If this conduit contains six #12 AWG THHN conductors, what size of conduit is required?

Since all conductors are the same size, refer to Table C8 of *NEC* Appendix C, scan down the left-hand column until THHN insulation is found, and then move one column to the right and find #12 AWG conductor. Note that the next column to the right lists ten #12 conductors in ½-in conduit. Since six conductors is all that is in this conduit run, and since ½ in is the smallest trade size conduit, this will be the size to use.

Installing Conduit

The normal installation of rigid metal conduit, intermediate metal conduit (IMC) and electric metallic tubing (EMT) requires the provision of many changes of direction in the conduit runs — ranging from simple offsets at the point of termination at outlet boxes and cabinets to complicated angular offsets at columns, beams, cornices, and the like.

Unless the contract specifications dictate otherwise, such changes in direction are accomplished, particularly in the case of smaller sizes, by bending the conduit or tubing as required. In the case of 1¼-in and larger sizes, right-angle changes of direction are sometimes accomplished with the use of "factory" elbows or conduit bodies. In most cases, however, such changes in direction are accomplished more economically by making conduit bends in the field.

Another good reason for making on-the-job bends is when multiple runs of the larger conduit sizes are installed. Truer parallel alignment of multiple runs is maintained by using on-the-job conduit bends rather than using factory elbows. Such bends can all be made from the same center, using the bend of the largest conduit in the run as the pattern for all other bends.

Since many raceway systems are run exposed, learning to install neat-looking conduit systems in an efficient and workmanlike manner is the basic trademark of a good electrician. Every electrician working on commercial and industrial electrical installations must therefore learn how to calculate and fabricate conduit bends — with both hand and power conduit benders, and take pride in performing the best work possible.

NEC Requirements

In general, the *NEC* requires that metal conduit bends must be made so that the conduit will not be damaged during the operation, and that the internal diameter of the conduit will not be effectively reduced in size. To accomplish this, the *NEC* further specifies the minimum radius of the inner-edge curve of a conduit bend which, in general, requires that the inside radius of an elbow must be at least six times the internal diameter of the conduit when conductors without lead sheath are to be installed. *See* Figure 5-3. There is good reason for this rule. When the inside radius of an elbow is less than six times the inside diameter, wire pulling becomes extremely difficult and the insulation on the conductors may be damaged.

When conductors with lead sheath are to be installed, the inside radius must be increased according to *NEC* Table 346-10.

The *NEC* further states that no more than four quarter bends (360 degrees total) may be made in any

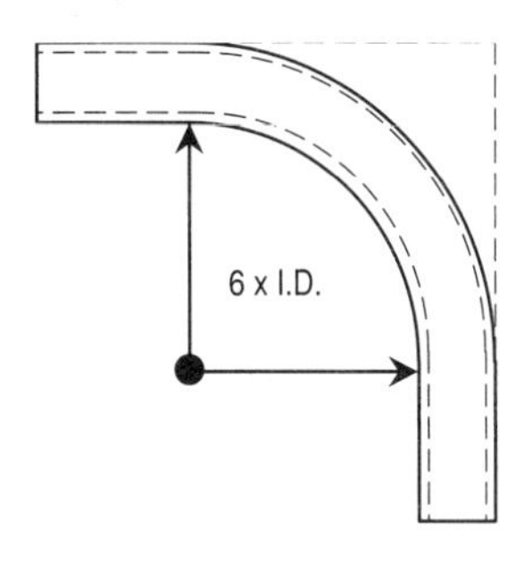

Figure 5-3: Inside radius requirements.

one conduit run between boxes, cabinets, panels, or junction boxes; that is, between pull points.

Some electricians are prone to consider that offsets, kicks and saddles are not bends — especially in areas where the electrical inspectors are lax. These electricians count only those bends that are actually a quarter circle (90 degrees). The misconception of this is quickly apparent when wires are pulled in. Offsets and saddles add just as much resistance to pulling in conductors as any 90-degree elbow. An offset, for example, takes two 45-degree bends which equal one 90-degree bend. A "saddle" should be counted as two quarter bends of 180 degrees.

A 15-degree "kick" in a conduit run may seem insignificant, but after several dozens of these kicks are incorporated into the run, the difficulty of pulling wire becomes apparent. The number of degrees in each kick should be included in the total count, and in no case should the total number (number of bends × number of degrees in each bend) exceed 360 degrees. This means that a maximum of 24 15-degree kicks may be incorporated into one conduit run; 12 kicks and two 90-degree bends; 12 kicks and two offsets; 12 kicks and one saddle, etc. These are the maximum number allowed. Many electricians prefer to install pull boxes at closer intervals to reduce the number of bends — especially when the larger conductor sizes are being pulled. The additional cost of the pull boxes and the labor to install them is often offset by the labor saved in pulling the conductors. A lesser number of bends also offers greater insurance against damaging the conductors during the pull.

Types Of Bends

Several types of bends are used in most conduit systems. A brief description of each follows:

Elbow: An elbow or "ell" is a 90-degree bend that is used when a conduit must turn at a 90-degree angle. In single conduit runs, when the larger sizes of conduit are being installed, factory elbows are frequently used to save labor on setting up a power bending machine. However, in multiple conduit runs, a neater job will result if on-the-job "sweep" or concentric bends are properly calculated, and installed as shown in Figure 5-4.

Offset: An offset consists of two 45-degree bends and is used when the conduit run must run over, under, or around an obstacle. An offset is also used at outlet boxes, cabinets, panelboards, and pull boxes as shown in Figure 5-5.

Kick: A kick is a minor change in direction of a conduit run. It is used mostly where the conduit run will be concealed as in "deck work." The first bend in an offset, for example, is really a "kick" as shown in

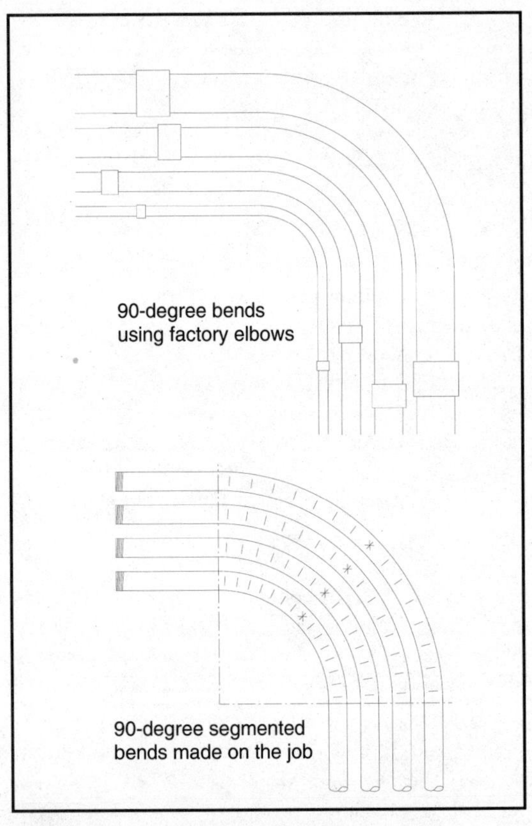

Figure 5-4: Comparison of 90-degree bends — factory and made-on-the-job bends.

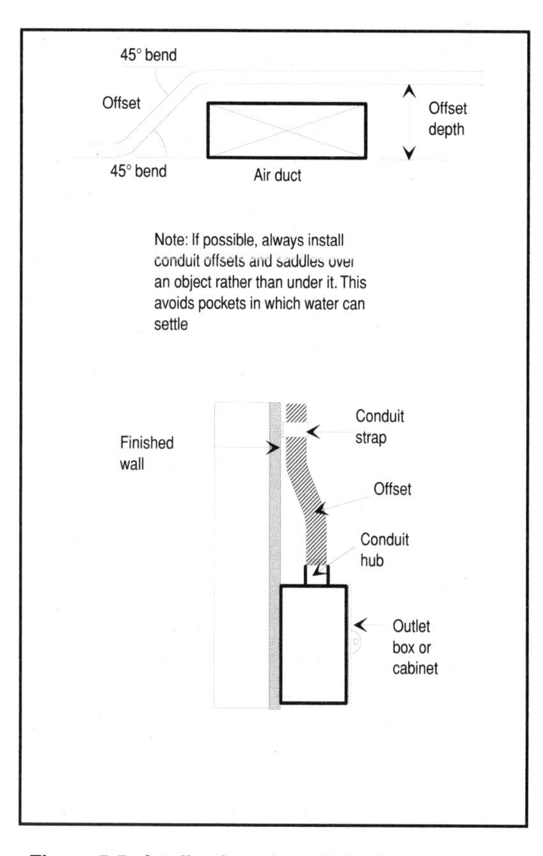

Figure 5-5: Application of conduit offsets.

Figure 5-6; another kick (in the opposite direction), however, transforms the kick into an offset.

Saddle: A saddle is used to cross a small obstruction or other runs of conduit. A saddle is made by marking the conduit at a point where the saddle is required, and placing a bender a few inches ahead of this point. Bends are made as shown in Figure 5-7 in approximately 30-degree increments. In most cases, the bends should be as close together as the bender will permit. If the obstruction is very large, use two offsets instead.

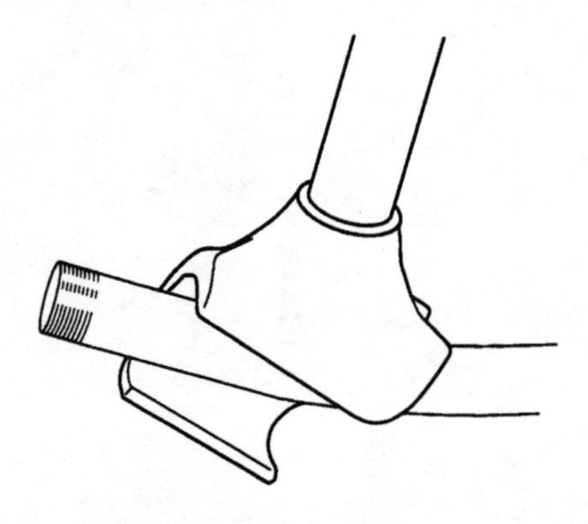

Figure 5-6: A kick in a conduit run is a minor change in direction.

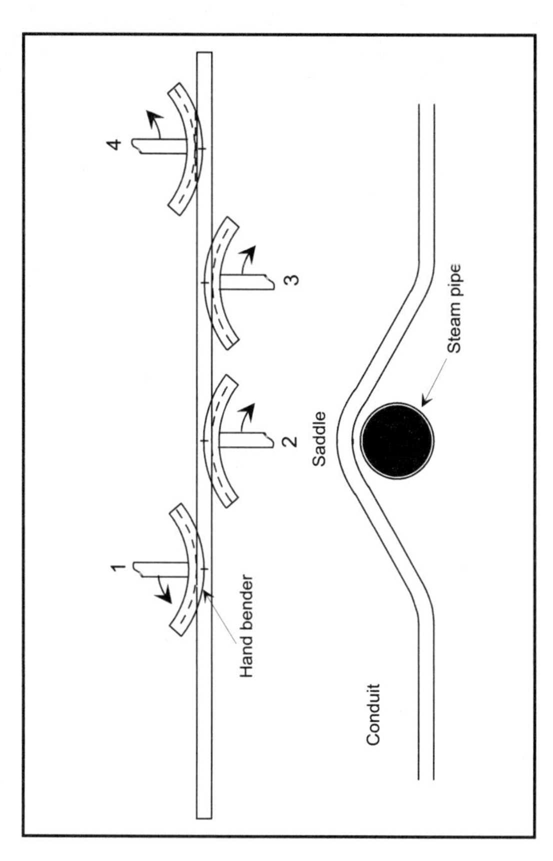

Figure 5-7: Practical application of a saddle bend.

Bending Conduit

Figure 5-8 shows a typical hand bender. Hand benders are convenient to use on the job because they are portable and no electrical power is required. Such benders have a shape that supports the walls of the conduit being bent. The hand bender is used for bending EMT, rigid conduit, and IMC. Hand benders provide a bending radius that conforms to the *NEC* for any size of conduit.

Most hand benders are designed to bend rigid conduit and EMT of corresponding sizes. That is, a single hand bender can bend either ¾-in EMT or ½-in rigid conduit. This is because the corresponding sizes of conduit have nearly equal outside diameters.

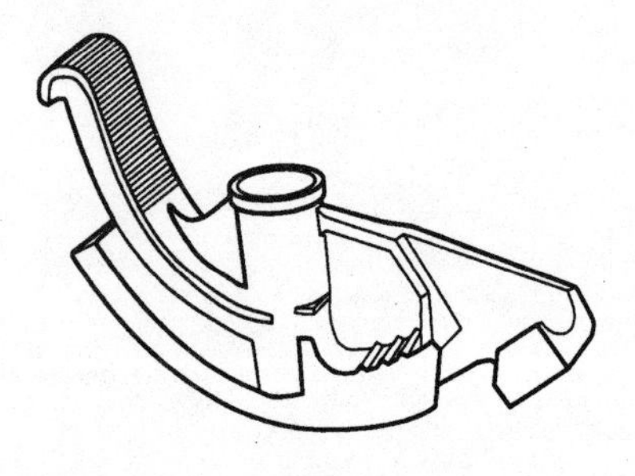

Figure 5-8: Typical hand bender.

There are two basic ways to use a hand bender. One way is to place the conduit on the floor and apply pressure to the bender with one foot on the step of the bender; then pull the handle toward yourself. A second method is for saddle or offset bends. The handle of the bender is placed on the floor and braced between one foot and the knee. Pressure is then applied to the conduit with the hand to form the bend.

A *hickey* (Figure 5-6) should not be confused with a hand bender (figure 5-8). The hickey is designed for bending small sizes of rigid conduit only because very little support is given to the walls of the conduit being bent. Consequently, the hickey functions quite differently from the hand bender.

When a hickey is used to bend conduit, both the bend and radius must be formed simultaneously. In doing so, care must be taken so as not to flatten or kink the conduit.

To use a hickey to bend rigid metal conduit, first make a small bend of approximately 10 degrees. Then the hickey is moved to a new position on the conduit and another small bend is made; again, not more than 10 degrees. This process is continued until the bend is completed. In the hands of an experienced electrician, the hickey is an excellent tool. It can be used to effectively stub-up conduits in slabs and decks.

Another bending tool that is frequently used on commercial projects is a special-purpose device

Figure 5-9: Little Kicker being used to make offsets in EMT.

called the "Little Kicker." This tool is designed to produce offsets in EMT with one motion of the tool's handle. Two models are available: one for ½-in EMT and the other for ¾-in EMT. The Little Kicker is an excellent time-saving device for making large numbers of offsets in the smaller sizes of EMT for terminating into junction or outlet boxes. *See* Figure 5-9.

Geometry Of Conduit Bending

Bending conduit requires some knowledge of basic geometry. You are probably already familiar with most of the concepts needed, but a brief review should prove helpful.

Right Triangle: A right triangle is defined as any triangle with one 90-degree angle. The side directly opposite the 90-degree angle is called the hypotenuse and the side on which the triangle sits is the base. The vertical side is called the height. On the job, the right triangle is applied when making offset bends; that is, the offset forms the hypotenuse of a right triangle as shown in Figure 5-10.

Circle: A circle is defined as a closed curved line whose points are all the same distance from its center. The distance from the center point to the edge of the circle is called the radius. The length from one edge of the circuit to the other edge is the diameter. The distance around the circle is called the circumference. A circle can be divided into four equal quadrants. Each quadrant accounts for 90 degrees, making a total of 360 degrees. When a 90-degree conduit bend is made, one-fourth of a circle (one quadrant) is used.

Concentric circles are several circles that have a common center but the radius of each circle is different. The concept of concentric circles can be applied to concentric 90-degree bends in conduit. Such bends have the same center point, but the radius of each is different. This relationship is shown in Figure 5-11. Also refer to Figure 5-4 to review the appearance of actual concentric sweep conduit bends.

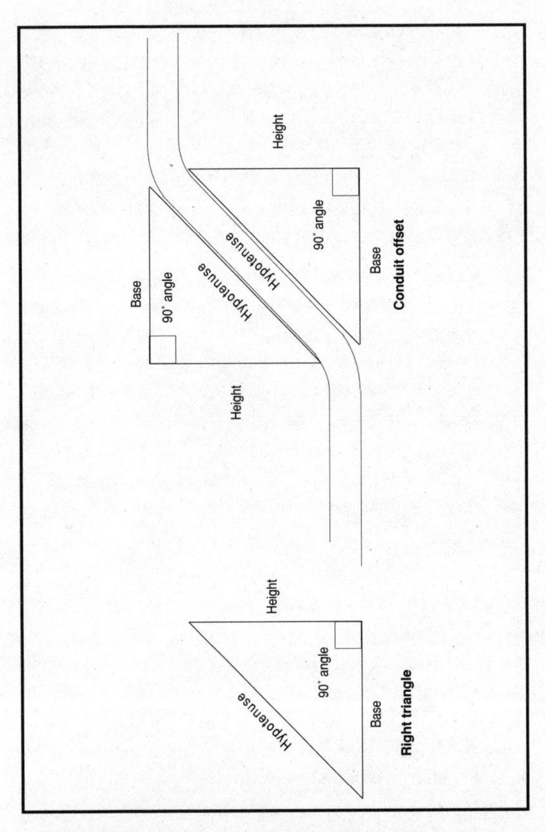

Figure 5-10: Right triangle and the relationship to an offset bend.

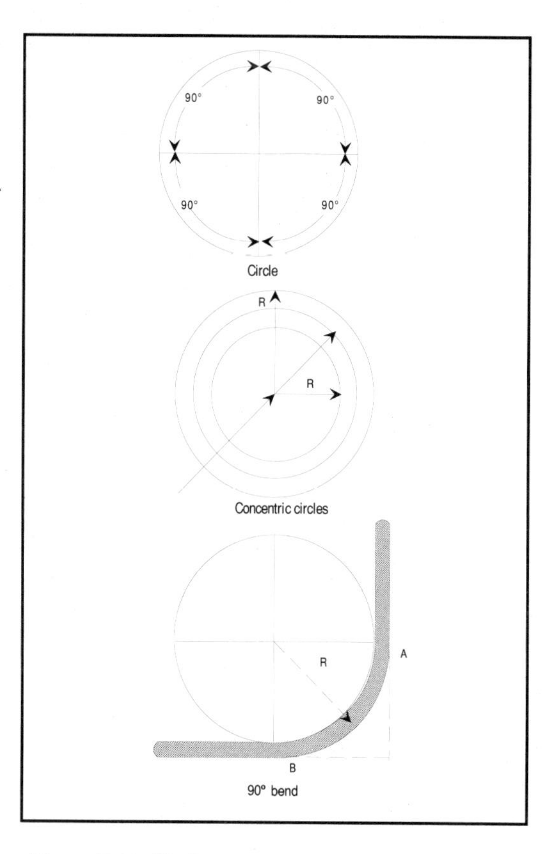

Figure 5-11: Circles and 90-degree conduit bends.

To calculate the circumference of a circle, use the equation:

$$C = \pi D$$

In this equation, C = circumference, D = diameter, and $\pi = 3.14$.

Another way of stating the equation for circumference is:

$$C = \pi 2R$$

In this latter equation, R = the radius or ½ the diameter.

To calculate the arc of a quadrant use the following equation:

$$\text{Length of arc} = (.25)2\pi R = 1.57R$$

In other words, the arc of a quadrant equals ¼ the circumference of the circle. Therefore, the arc of a true 90-degree conduit bend also equals ¼ the circumference of a circle; four 90-degree conduit bends equals a full circle or 360 degrees.

Making 90-Degree Bends

The 90-degree stub-up bend is probably the most basic bend of all, especially on deck jobs for stubbing up conduit for outlet boxes prior to the concrete pour, or when the conduit must make a 90-degree change in direction. Before beginning to make the bend, two measurements must be known:

- The desired rise or height of the stub-up
- The take-up distance of the bender

"Take-up" is the amount of conduit the bender will use to form the bend. Take-up distances are usually listed in the bender manufacturer's instruction manual; sometimes these figures are inscribed directly on

EMT	Rigid Conduit	Take-Up
½"	—	5"
¾"	½	6"
1"	¾	8"
1¼"	1"	11"

Figure 5-12: Typical conduit bender take-up distances.

the side of the bender. A sample table is shown in Figure 5-12.

Once the take-up has been determined, subtract it from the stub-up height. Measure back from the end of the conduit and mark that distance on the conduit with a felt-tip marker; make the mark all the way around the conduit. The mark will indicate the point at which you will begin to bend the conduit. Line up the starting point on the conduit (the mark just made) with the starting point on the bender. The latter is usually in the form of an arrowhead or other mark on the side of the bender.

Once you have lined up the bender with the mark that you made on the conduit, use one foot to hold the conduit steady, keeping the heel of this foot on the floor for balance. Apply pressure on the bender foot pedal with your other foot. Make sure you hold the bender handle level and as far up on the handle as possible to get maximum leverage. Then bend the conduit in one smooth motion, pulling as evenly as possible.

When using the take-up method to bend conduit, the bender is always placed on the conduit and the bend is made facing the end of the conduit from which the measurements were taken. *See* Figure 5-13 for the measurements and bender placement on ½-in conduit.

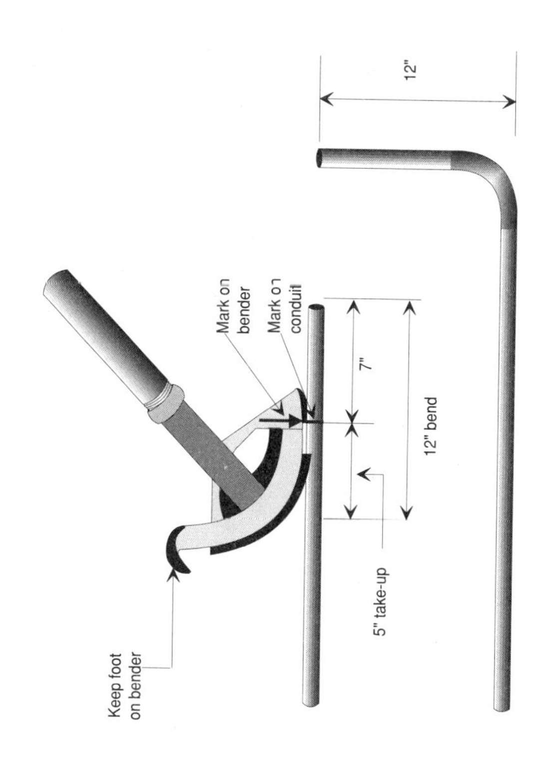

Figure 5-13: Measurements used to make a 12-inch stub-up on ½-inch conduit.

After finishing the bend, check to make sure that the correct angle and measurements were made. Use the following steps to check a 90-degree bend:

Step 1. With the conduit on the floor, in the same position that it was while the bend was being made, measure from the floor to the top of the stub-up to make sure it is the correct height.

Step 2. With the back of the bend on the floor, measure to the end of the conduit to make sure it is the right length.

Step 3. Check the 90-degree angle of the bend with a square or at the angle formed by the floor and a wall. You may also use a magnetic torpedo level.

If the conduit is slightly overbent past the desired angle, use the bender to bend the conduit back to the correct angle.

The method just described will produce a 90-degree "one-shot" conduit bend; that is, it took one motion of the bender to form the bend. A segment bend is any bend that is formed by a series of bends of a few degrees each. A "shot" is actually one bend in a segment bend.

Bender Take-Up: Hand benders are available in four sizes:

- ½ in
- ¾ in
- 1 in
- 1¼ in

Each size has a definite "take-up" or "rise." When a full 90-degree bend is made from the floor, the end of the arc will be 5 in off the floor with ½-in EMT, 6 in off the floor with ¾-in EMT, and 8 in off the floor with 1-in EMT.

When working on the floor, a full 90-degree bend is easily made by pulling the bender handle toward the operator until the end of the conduit is pulled from its horizontal position to a vertical position. Maintain constant pressure on the step of the bender with your foot to prevent kinking the conduit. A 45-degree bend is made by pulling only until the bender handle points straight up.

Gain: The gain is the distance saved by the arc of a 90-degree bend. Knowing the gain can help to precut, ream, and prethread both ends of the conduit before the bend is made. This will make the work go quicker because it is easier to work with conduit while it is straight. Figure 5-14 shows that the overall length of a piece of conduit with a 90-

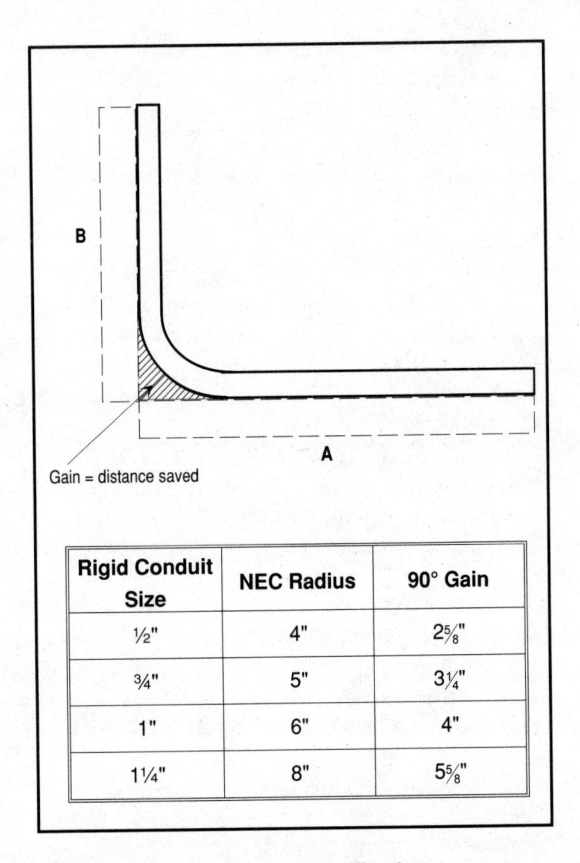

Rigid Conduit Size	NEC Radius	90° Gain
½"	4"	2⅝"
¾"	5"	3¼"
1"	6"	4"
1¼"	8"	5⅝"

Figure 5-14: Conduit gain specifications.

degree bend is less than the sum of the horizontal and vertical distances, when measured square to the corner.

Developed length = (A + B) - Gain

A sample conduit-bender manufacturer's gain table is also shown. These tables are short-cut devices to calculate the gain for any conduit size and will prove useful on all jobs.

Back-To-Back Bends

Back-to-back or "U" bends consist of two 90-degree bends, placed back-to-back as shown in Figure 5-15.

To make back-to-back bends, first bend a conventional 90-degree bend as discussed previously and labeled "X" in Figure 5-15. To make the second bend, measure the required distance between the bends from the back of the first bend. This distance is labeled "L" in Figure 5-15. Reverse the bender on the length of conduit and place the bender's "back-to-back" indicating mark at point "Y" on the conduit. Note that the measurements from point "X" to point "Y" are taken from the outside edges of the conduit. Now, holding the bender in the reverse position and properly aligned, apply foot pressure and complete the second bend.

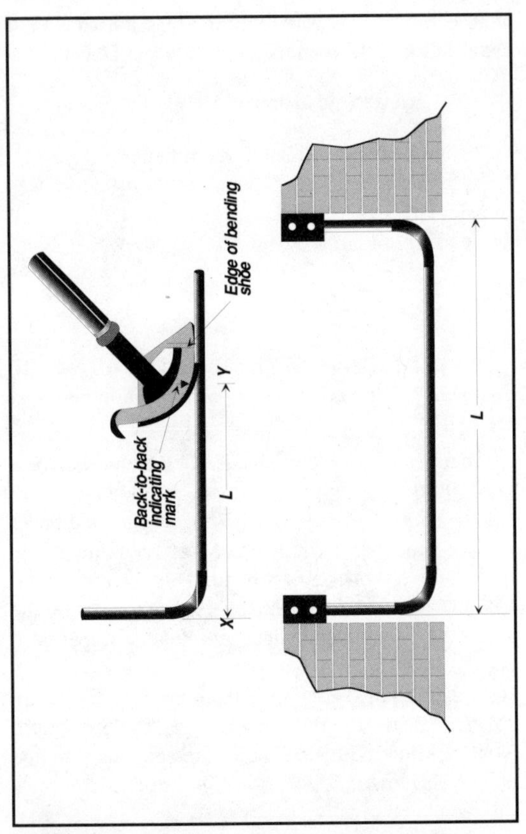

Figure 5-15: Process of making back-to-back bends.

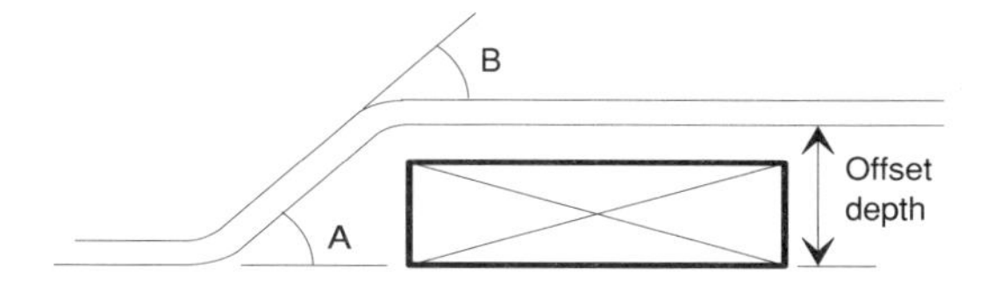

Figure 5-16: Typical offset.

Offsets

Many situations require that conduit be bent so that it can pass over objects such as beams and other conduits, or for entering meter cabinets, panelboards, and junction boxes. Bends used for this purpose are called "offsets." To produce an offset, two equal bends of 90-degrees or less are required, a specified distance apart as shown in Figure 5-16.

In making conduit offsets, be aware that the degree of bend dictates the space requirements for the total offset; that is, the smaller the angle of bend, the larger the space needed to complete the change in elevation. If space is not a problem, it is best to keep the angle of the bends small so that pulling the wire will be easier.

Offset Depth in Inches	Distance Between Bends	Angle of Bends	Conduit Length-Lossin Inches
1"	6"	10°	1/16"
2"	5 1/4"	22 1/2°	3/8"
3"	6"	30°	3/4"
4"	8"	30°	1"
5"	7"	45°	1 7/8"
6"	8 1/2"	45°	2 1/4"
7"	9 3/4"	45°	2 5/8"
8"	11 1/4"	45°	3"
9"	12 1/2"	45°	3 3/8"
10"	14"	45°	3 3/4"

Figure 5-17: Chart of measurements for making conduit offsets.

Angle	Multiplier
10°	5.76
15°	3.86
22.5°	2.61
30°	2.00
45°	1.41

Figure 5-18: Offset multipliers for various angles.

Once the angle of the bends and the required height of the offset have been determined, the distance between the bends may be calculated. Bender manufacturers typically provide guides that can help you determine this distance at a glance. A sample chart appears in Figure 5-17.

Charts are also available that provide a multiplier for each angle used in conduit offsets. One such table is shown in Figure 5-18. The distance between bends is equal to the height of the offset times the multiplier for the angle used.

For example, a bend of 45-degrees with a height of 10 in requires 14 in between the first and second bend. So, in this case, you would place indicating marks on the conduit 14 in apart to indicate where the first and second bends should start.

Once the angle of bend is known along with the height of the offset and the distance between bends, the location of the first bend can be determined. Keep in mind that the conduit will "shrink" a certain amount per inch of offset depth. Using the chart in Figure 5-17, it can be seen that the amount of conduit will shrink 3¾ in for the example cited above. This means that the bender should be placed 3¾ in ahead of the mark for the first bend. Remember that most benders have an arrow or other indicating mark that shows where to place the bender for various types of bends.

To continue, make the bend at the first mark and then make the bend at the second mark, keeping the conduit running in the same direction through the bender's shoe.

It is important to be precise when measuring and bending offsets. A few degrees too much will make the offset too high, and bending less than the required number of degrees will make the offset too low. By the same token, rounding off numbers too much during calculations will result in a length of conduit that won't fit.

Another important consideration in making offset bends is to make certain that the bends are precisely in line with each other. After making the bends, lay the length of conduit on the floor to test for alignment.

Saddle Bends

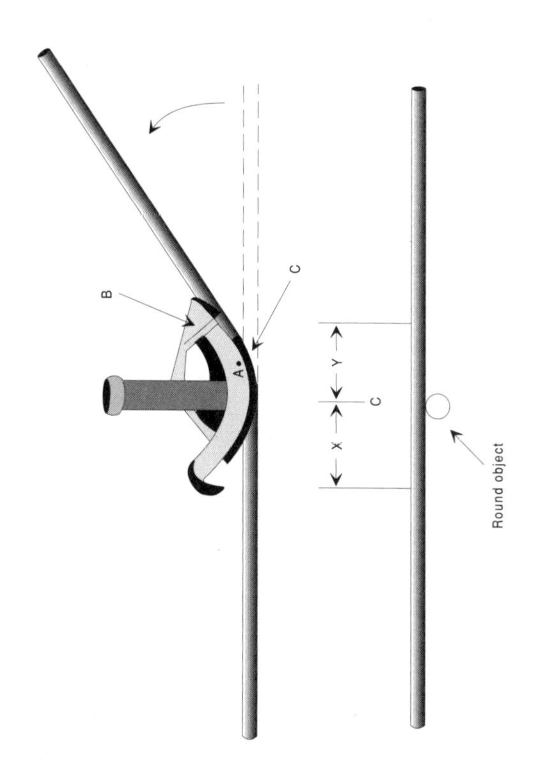

Figure 5-19: Steps in making a saddle bend.

A saddle bend is a series of bends that are used to change elevation to clear an obstruction and then return to the original elevation. You can think of a saddle as a "double offset." To bend a saddle, place a length of conduit across the object to be saddled as shown in Figure 5-19.

Point "C" will be the center of the finished saddle. Calculate twice the diameter of the object to be saddled. Lay out "X" that distance from "C." "Y" is the same distance in the opposite direction from "C."

Place the conduit in the bender so that "C" on the conduit is at the indicating notch on the bender and make a 45-degree bend. Remember, a 45-degree bend is reached when the bender handle is straight up. Rotate the conduit 180 degrees in the bender, place the "X" on the conduit at "B" on the bender and make a return bend of 22½ degrees. Duplicate this procedure by placing "Y" on the conduit at "B" on the bender.

Mechanical Benders

Conduit bends are normally made in the smaller sizes of conduit and tubing by hand with the use of hickeys or EMT bending tools as discussed previously. However, on many projects, an advantage can be gained by the use of mechanical bending equipment with suitable adjustable stops and guides.

With the use of such equipment, the exact bend can be duplicated in quantity with a minimum of effort. The angle of bend and the location of the bend in relation to the end of the length of conduit is preset.

One popular mechanical bender is shown in Figure 5-20. This type of bender was originally called "Chicago bender," as it was originally invented and manufactured by the Chicago Equipment and Manufacturing Company. Today, however, this type of bender is manufactured by several different companies and a more correct name may be, "portable mechanical conduit bender." In any event, you will probably still hear the term "Chicago bender" on many jobs — especially from the "old-timers."

The Greenlee Model 1818 mechanical bender is very popular with many electrical contractors. In use, a length of conduit is placed in position and secured in place; a long bending handle is then pulled around and the bend completed. This type bender may be used as a one-shot bender for the smaller sizes of conduit (bypassing the ratchet mechanism). The ratchet mechanism, however, is usually activated when bending the larger sizes of conduit to make the work easier. The Greenlee Model 1818 bending tool is suitable for making bends in conduit sizes up to 2 in EMT, 1¼ in IMC, and 1½ in rigid conduit, provided the proper bending accessories are used; that is, bending shoes, follower bars, etc. — all of which warrant added discussion.

Figure 5-20: Greenlee ratchet bender.

Bending shoes and follower bars are designed to form a particular radius bend for a certain type and size of conduit; that is, EMT, IMC or rigid conduit. These accessories should be treated as precision instruments; treated otherwise will result in inaccurate bends, kinks, and the like. The first consideration is to use only the proper shoe and follower bar for the type and size of conduit being bent. For example, never use an EMT shoe for bending rigid conduit, or a 3-in shoe for bending, say, 2½-in conduit. In general, make certain that the bending shoes and follower bar are compatible with the type of conduit to be bent. To do otherwise will result in damage to the tool and inaccurate bends.

The ratchet feature is normally engaged for the larger sizes of conduit, whereas a spring-loaded pawl engages the ratchet for easier bending in segments. For the smaller sizes of conduit, however, the ratchet may be bypassed so that the bend can be made in one shot.

A bending gauge (Figure 5-21) with an adjustable pointer on the bender is helpful when making multiple bends at the same angle. This pointer is set at the desired angle and then the set screw is tightened. As the bend is being made, the handle is operated until the pointer reaches the index mark. To ensure the correct angle of bend, the first bend should be checked with a bending protractor (Figure 5-21) and any necessary adjustments made to the bending-gauge

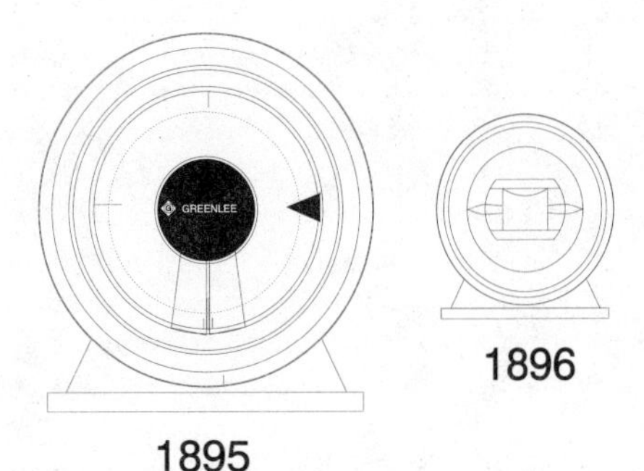

Figure 5-21: A bending protractor is used to obtain bends to an exact angle.

pointer before continuing. All successive bends will be exactly as the first.

Mechanical Stub-Ups

Stub-ups are quickly and easily made with mechanical benders. A "deduct" decal is provided on all Greenlee benders, but sometimes these decals become damaged, making them difficult to read. Therefore, back-up charts should be provided on all jobs. With this "deduct" chart on hand, the following is an example of making a 90-degree stub-up to a given height.

Let's assume that you're working on a "deck" job and need a number of one-inch rigid stub-ups with a rise of 15 in each. Check the "deduct" chart on the bender for a 15 in stub-up using 1 in rigid conduit. Note that 11 in should be deducted from the total rise of 15 in. Consequently, 15 – 11 in = 4 in. Measure back from the end of the conduit this amount (4 in) and make a mark. Encircle the entire conduit at this point so you won't lose the mark once the conduit is placed in the bender. Most electricians like to use a black felt-tip marker for marking conduit.

Load the conduit into the bender with the mark lined up with the front of the bender hook. Engage the ratchet and start "pumping" the bender handle until the bender pointer reaches the preset index mark for 90 degrees. Move the bender handle forward, remove the conduit from the bender, and check its height. It should be exactly 15 in. If the height of the bend is slightly off, make the necessary adjustments before continuing. Once the correct height is reached, the remaining bends will also be correct.

Mechanical Offsets

The decal chart on the bender that provides "deduct" information for stub-ups also contains data for making offsets — using 20-, 30- and 45-degree bends. This chart is necessary to make perfect offsets every time with the mechanical bender.

Let's assume that you're running a raceway system with ½-in rigid conduit and an air duct must be bypassed, requiring the conduit run to be offset. After taking measurements on the job, we find that an offset of 12 in is needed to clear the air duct.

Measure the distance from the end of the conduit to the start of the first bend; mark the conduit as before. Referring to the chart on the bender with offset information, we decide to make the offset with two 45-degree bends. The chart indicates that the distance between bends is 16$\frac{5}{16}$ in. Therefore, measure and mark this distance back from the first mark.

Insert the conduit into the bender and line up the first mark with the front of the bender hook. The ratchet may be used, but for ½ in conduit, the ratchet override on the front of the bender shoe is normally employed. Make the first 45-degree bend. Move the bender handle forward to release the conduit. Now slide the conduit forward through the bender hook until the second mark lines up with the front of the bender hook and then turn the bend over so the end of the conduit is pointing downward—toward the deck. Also make sure that the first bend lines up with the next bend that will be made to prevent a wow (a crooked bend) in the conduit. Once everything is aligned, engage the bender handle and make another 45-degree bend. The height of the offset should be exactly 12 in. *See* Figure 5-22.

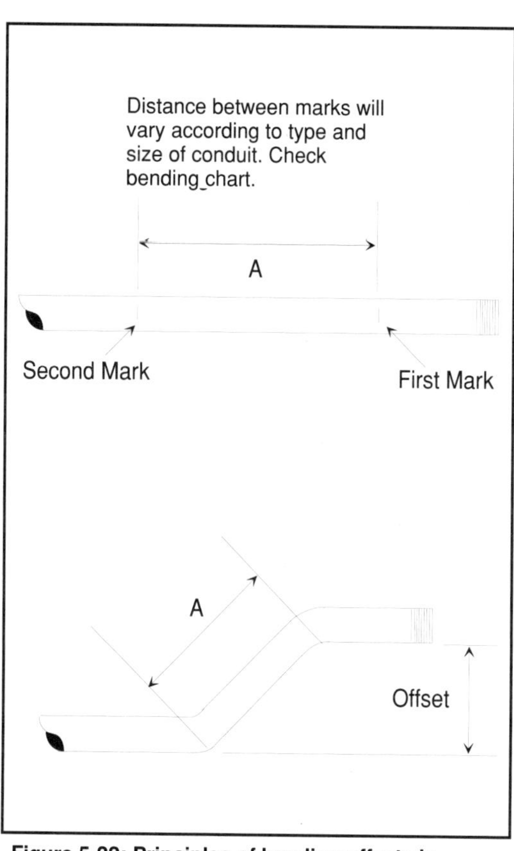

Figure 5-22: Principles of bending offsets in conduit.

The distances between marks for offset bends will vary, depending upon the size and type of conduit being bent, so always check the offset information on the bender.

ELECTRIC AND HYDRAULIC CONDUIT BENDERS

Bends in larger sizes of conduit (over 2 in EMT, 1¼ in IMC, or 1½ in rigid conduit) are normally made with the use of hydraulic benders of different types. In some instances, so-called bending tables have been developed for use with bending tools or hydraulic benders to simplify making bends to certain dimensions. Various adaptations of benders have been developed to serve certain specific purposes. For example, one type of hydraulic bender is designed for making bends in a section of conduit which has been installed in a raceway system.

Electric Conduit Benders

Electric conduit benders operate on the same basic principle as the mechanical benders described previously except that the bending is accomplished by a gear motor rather than manual power. There are several types of electric benders on the market, but you will probably see more of the Greenlee 555 than any other (*See* Figure 5-23). This bender will make stub-

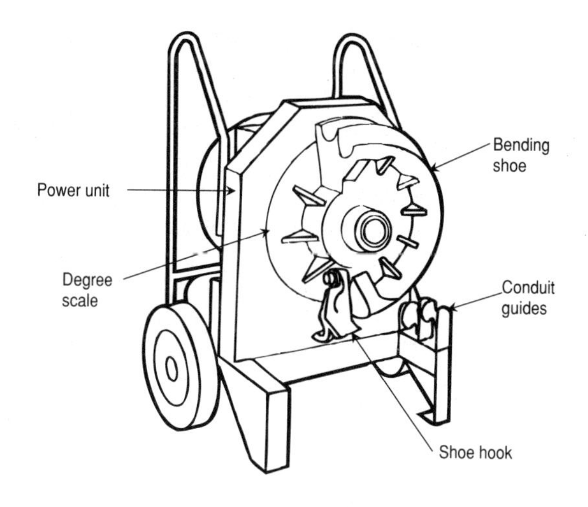

Figure 5-23: Greenlee 555 electric bender.

ups, offsets, and saddles quickly and easily at any location where a 120-V receptacle is available. Furthermore, this bender will make 180° one-shot bends in ½- through 2-in EMT, IMC or rigid conduit.

Bending charts are in the form of decals attached to the bender for quick reference on the job. These include "deduct" and "springback" figures along with information for making offsets.

The Greenlee 555 comes as a power unit and the bending shoes for various types of conduit are sold as accessories. These shoes quickly snap into place on the power unit, ready for use in seconds.

To describe the operation of the "Five Fifty-Five," let's assume that we need a 16-in stub-up in a piece of 1¼ in rigid conduit. Measure and mark a piece of conduit 16 in from one end. Check the "deduct" chart on the bender for this size conduit; the figure is 12¾ in. Therefore, 12¾ in must be deducted from 16 in which leaves 3¼ in. So a mark is made this distance from the end of the conduit. Encircle the conduit with this mark so it will be plainly visible during the bend.

Insert the conduit in the bender with the mark lined up with the front of the bender hook. For this particular bend, the mark is 3¼ in from the end of the conduit.

The three-position operating switch on the 555 bender is attached to a flexible cord. The center position de-energizes the machine; the up position (forward) is for bending, and the down position (reverse) is for unloading the conduit after the bend has been made. When released, the switch automatically springs to the center or off position. A pointer on the shoe indicates the degree of bend as the bend is being made.

The machine is "jogged" or "inched up" until the shoe pointer lines up with the zero mark on the degree scale. The switch is then pressed upward and held in this position to start the bend. As the pointer approaches the 90-degree mark, refer to the "springback" chart for the size and type of conduit being bent. In this case (1¼ in rigid), the chart indicates 95°. Therefore, the pointer should pass the 90-degree mark and stop at 95° to allow for springback when the conduit is removed from the bender. Press the operating switch to the down position and unload the conduit from the bender; the stub-up is ready for installation.

An offset is made in the 555 similar to the method described for the mechanical bender; that is, the first mark is located on the conduit and then offset information is obtained from the bender chart. In the case of a 16-in offset in a length of 1¼ in rigid conduit, the chart indicates a distance of 22⅝ in between marks for a 16 in offset using 45° bends.

Mark the conduit and insert the conduit into the bending shoe, positioning the mark even with the front of the bender hook. Start the bend as discussed previously until the pointer reaches the 45-degree mark on the scale; allow for any springback as indicated in the chart. Reverse the motor until the conduit is loose, and then turn the conduit upside down. Position the second mark at the front of the bender hook, making certain the conduit is aligned to prevent a wow in the bend. Start the second bend until the pointer reaches the 45-degree mark, reverse the bender and unload the conduit. A magnetic torpedo level placed on the side of the conduit will help align bends for offsets.

Finally, check the bends for accuracy. Although rebending is possible, it is not a good practice. Rebending puts considerable strain on the conduit and while it may not break, the coating may crack and cause corrosion.

Speed Benders

The Greenlee 555 Speed Bender operates basically the same as the standard 555 except the Speed Bender utilizes remote digital control with easy-to-use bending charts and instructions to insure fast, accurate and consistent bends in conduit sizes from ½ to 2 in.

This bender can be operated upright or laid on its back for large offsets and saddles. In operation, rather than holding the switch up in the bend position as with the standard 555, the operator sets the bend wanted via digital controls and the bender automatically bends to that degree, once the conduit is placed in the bending shoe and the start control is activated. Otherwise, the conduit is marked and bent the same as described for the standard 555.

Hydraulic Conduit Benders

There are several types of hydraulic benders used on electrical construction jobs. In general, these benders make use of hydraulics to make conduit bends of various types. Most of these benders utilize lightweight aluminum components and a pin system that speeds up assembly and disassembly. Two basic types are in general use: the flip-top variety such as Greenlee's 882 and 881 series and the A-frame benders such as the Greenlee 777, 880, 884, 885, etc. series. *See* Figure 5-24. Basic operating principles are described in the paragraphs to follow.

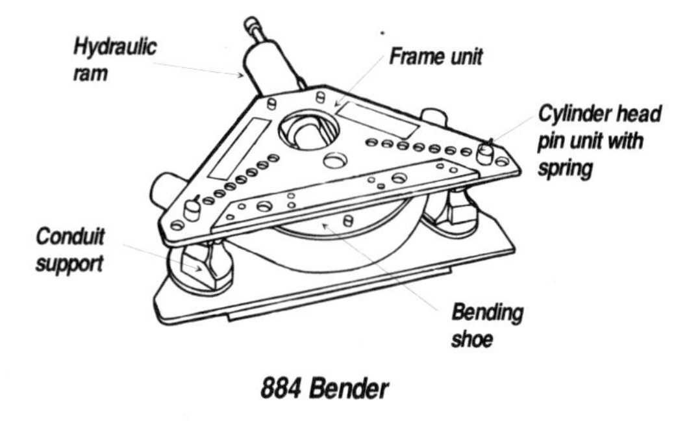

Figure 5-24: One common type of hydraulic conduit bender.

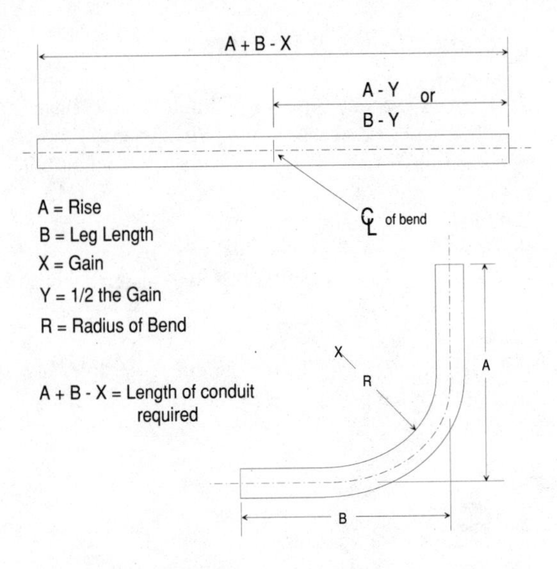

Figure 5-25: Principles of laying out stub-ups.

One-Shot Bending

Accurate one-shot stub-ups are easily made on hydraulic benders by applying a little basic geometry in making calculations, and then knowing the operating principles of the bender. For example, Figure 5-25 shows the reference points of a common 90-degree bend. Use the following procedure for laying out accurate stub-ups:

- Determine lengths A and B.
- Add length A and B, subtract X for length of pipe required.
- Subtract Y from length A or B to get center of bend.

The chart in Figure 5-26 gives useful information for making accurate stub-ups with Greenlee 777, 880, 883, and 884 hydraulic benders.

To make one-shot 90-degree bends, first determine the leg length and rise, the radius of the bend, gain, and half-gain (see Figure 5-25).

Calculate the developed length and the length of conduit required, if it is to be fitted between two existing conduits. The last step of the layout process is to determine the center of the bend. This can be done by taking the half-gain value from the distance A or B. Note the radii in the Radius of Bend column in the chart in Figure 5-26.

90-Degree Segment Bends

When bending conduit in segments with a hydraulic bender, the following factors must be determined:

- The size of conduit to be bent
- The radius of the bend

- The total number of degrees in the bend
- The developed length
- The gain of the bend

Pipe and Conduit Size	Radius of Bend R	Developed Length 90°	Gain X	½ Gain Y
½″	4″	6$\frac{5}{16}$″	1$\frac{11}{16}$″	⅞″
¾″	4½″	7$\frac{1}{16}$″	1$\frac{15}{16}$″	$\frac{15}{16}$″
1″	5¾″	9″	2½″	1¼″
1¼″	7¼″	11⅜″	3⅛″	1$\frac{9}{16}$″
1½″	8¼″	13″	3½″	1¾″
2″	9½″	14$\frac{15}{16}$″	4$\frac{1}{16}$″	2″
2½″	12½″	19⅝″	5⅜″	2$\frac{11}{16}$″
3″	15″	23$\frac{9}{16}$″	6$\frac{7}{16}$″	3$\frac{3}{16}$″
3½″	17½″	27½″	7½″	3¾″
4″	20″	31$\frac{7}{16}$″	8$\frac{9}{16}$″	4¼″

Figure 5-26: Dimensions of stub-ups for various sizes of conduit.

To determine the developed length for a 90-degree bend, multiply the radius by 1.57.

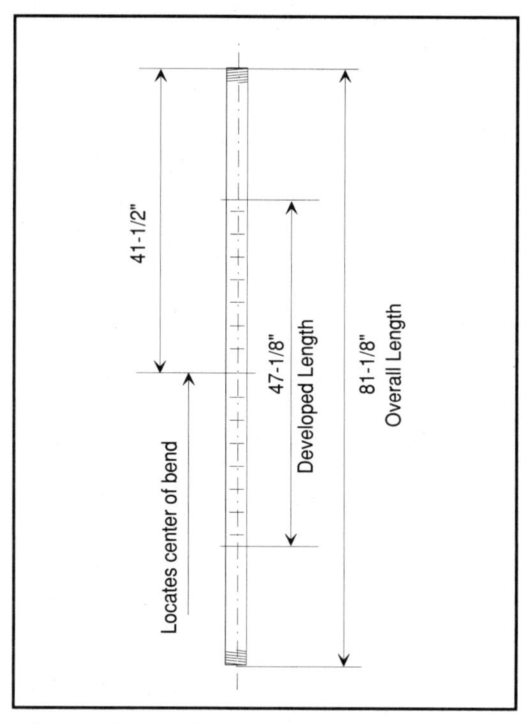

Figure 5-27: Principles of laying out segment bends on a 3-in conduit.

The next step is to locate the center of the bend. Most benders have the center mark indicated on the bending shoes. Once the center mark on the conduit is found, it is easy to locate the other bend marks (*see* Figure 5-27). You must now determine the number of shots that will make the bend to suit the requirements, preferably an odd number so that there are an equal number of bends on each side of the center mark. Next you calculate the width of the spaces for each segment bend and make the layout on the conduit. Make an equal number of spaces on each side of the center mark. The gain need only be determined if the bend is being fitted between two existing conduits or junction boxes.

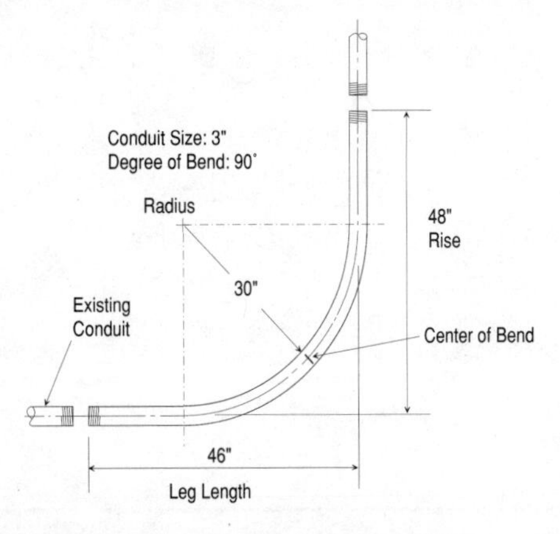

Figure 5-28: Specifications for sample bend.

To determine the bending data for a 90-degree bend using 3-in conduit with a rise of 48 in and a leg length of 46 in, and a centerline radius of 30 in, proceed as follows (*see* Figure 5-28):

1. Multiply the radius by 1.57 to determine the developed length. Therefore,

$$30'' \times 1.57 = 47.10'' \text{ developed length}$$

2. Determine the gain with a 90-degree bend where gain equals:

$$\text{Gain} = (2)(\text{radius}) - \text{Developed Length}$$

$$= (2)(30'') - 47\tfrac{1}{8}''$$

$$= 60'' - 47\tfrac{1}{8} = 12\tfrac{7}{8}''$$

3. To calculate the overall length of the conduit (OL), add the leg and stub-up length and subtract the gain. Refer to Figure 5-28.

$$\text{OL} = \text{leg length} + \text{rise} - \text{gain}$$

$$= 46'' + 48'' - 12\tfrac{7}{8}''$$

$$= 94'' - 12\tfrac{7}{8}'' = 81\tfrac{1}{8}''$$

4. Now locate the center of the required bend. Use the rise or stub-up dimension of 48 in. Subtract the radius (30 in), and add one-half of the developed length; that is,

$$\tfrac{1}{2}(47\tfrac{1}{8}) = 23.56 \text{ in}$$

$$= 23\tfrac{1}{2} \text{ in}$$

Therefore, the distance from the stub end to the center of the bend will be,

$$48'' - 30'' + 23\tfrac{1}{2}'' = 41\tfrac{1}{2}''$$

5. As a rule, six degrees or less per bend will produce a good bend for a 30-in radius. In this case, 6 degrees per bend will be used, making (90 ÷ 6 =) 15 segment bends. An odd number of segment bends is easy to lay out after finding the center mark, because there will be an equal number of spaces on each side of the center mark.

6. To determine the space between the segment marks, divide the developed length by the total number of segments; that is,

$47\frac{1}{8}$ or $47.125 \div 15 = 3.14''$

or about $3\frac{1}{8}''$ apart

7. Locate the conduit in the pipe holders — clamping them securely.

8. Place the center mark $41\frac{1}{2}$ in from one end of the conduit. Next mark seven points on each side of the center point, $3\frac{1}{8}$ in apart, for a total of 15 marks. These are the centers of the segment bends.

9. It is good practice to check the distance between the first and last bend marks, to be sure the layout is correct before starting the first bend. The distance from the first mark to the last is the developed length minus the length of one bend. (Actually you are subtracting one half of a segment bend from each end of the conduit.)

$$47\frac{1}{8}'' - 3\frac{1}{8}'' = 44''$$

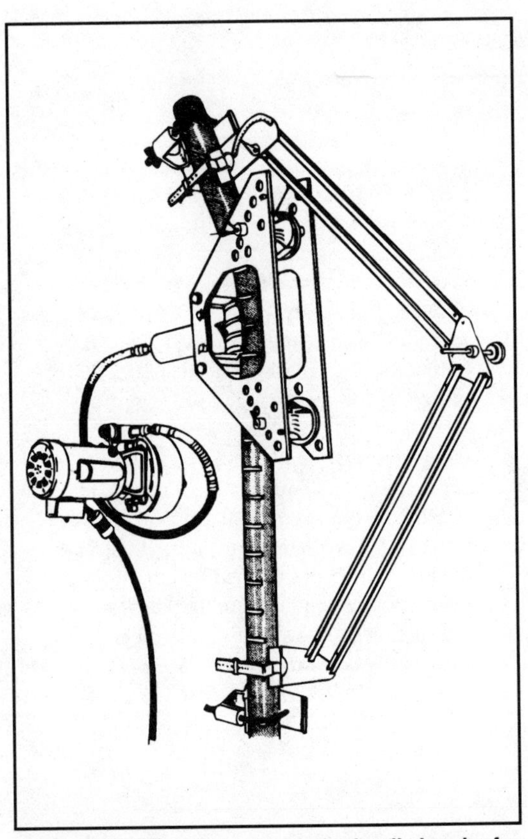

Figure 5-29: Conduit set-up in hydraulic bender for segment bends.

10. After locating the conduit in the bender, attach the pipe-bending degree indicator in a convenient location, as shown in Figure 5-29.

11. Attach the pipe supports with the proper face toward the conduit and insert the pipe support pins. Lock them in position by turning the small lock pin. Now proceed to make the series of bends.

12. Figure 5-29 shows the entire set-up. Begin by bending 6 degrees on the first mark. When this is done, the indicator will read 6 degrees. Release the pressure and check the springback; if any is found, over bend the same amount.

13. After the first bend, when using a bender with a rigid frame, move the pipe support one hole position in (toward the ram), that is, on the side that you have bent the conduit (*see* Figure 5-29).

14. Continue to bend to 12 degrees on the second mark. Check the springback. When the first bend in the conduit is moved past the one pipe support, the ram travel for the remaining bends will be exactly the same.

15. Follow this procedure until you get to the last mark, where you will be bending to 90 degrees. Stop at exactly 90 degrees, release the pressure, check the springback, and overbend the same amount. The result will be a 90-degree bend without any bows or twists.

Concentric Bending

If two or more parallel conduits must be bent in the same direction as in Figure 5-30, neater results will be obtained by using concentric bends. When laying out concentric bends, the bend for the innermost conduit is calculated first. In this example, the first bend has a radius of 20 in. If this dimension is multiplied by 1.57, it yields a value of 31$^{13}/_{32}$ in. This is the developed length of the shortest radius bend. To determine the radius of the second bend, you must add $^{1}/_{2}$ the outside diameter (OD) of the first conduit (1$^{3}/_{16}$ in), $^{1}/_{2}$ the outside diameter of the second conduit (1$^{3}/_{16}$ in), the distance between the two conduits (2 in), and the radius of the first bend (20 in). In this example, the radius of the second bend would be equal to 24$^{3}/_{8}$ in.

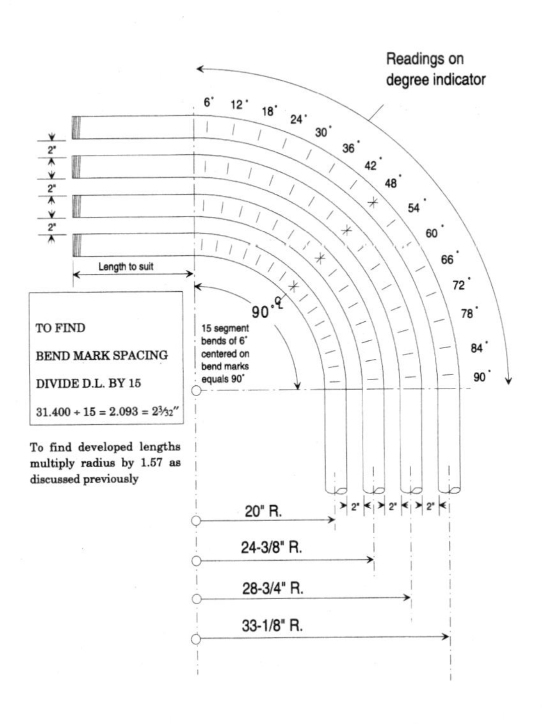

Figure 5-30: Principles of concentric bending.

The equation for the developed length (DL) would be:

$$1.57 \times R \text{ (radius)}$$

The developed length for the second bend is found by multiplying:

$$(1.57)(24.37) = 38\tfrac{1}{4}''$$

The radius and developed length of each successive bend are found in a similar manner.

After determining the developed length, you must establish the number of segment bends needed to form each 90-degree bend. In concentric bending, every bend must receive the same number of segment bends to maintain concentricity. As illustrated, 15 segment bends of 6 degrees each will total 90 degrees.

Note that the radius change from one bend to another affects the spacing of segment bends. To find the segment-bend spacing, divide the developed length of each bend by 15. For the first bend illustrated, the spacing for each segment bend is $2\tfrac{3}{32}$ in; for the second bend, $2\tfrac{9}{16}$ in and so on.

If the legs of the bends have to be a certain length, the gain must be considered just as in any segment-bending procedure. When conduits are not the same size, the radius of each successive bend can be found as follows:

- Determine the radius of the innermost bend.
- Calculate ½ the outside diameter of the innermost conduit and of the next adjacent conduit.
- Note the distancc bctwccn thcsc two conduits.
- Add these quantities.

Offset Bends

Many situations require a conduit to be bent so that it can pass by or over objects such as beams and other conduits, or enter panelboards and junction boxes. Bends used for this purpose are called "offsets." To produce an offset, two equal angle bends of less than 90 degrees are required, a specified distance apart. This distance is determined by the angle of the two bends and can be calculated by using the following procedure and the table in Figure 5-31 beginning on the next page.

First, determine the offset needed, then the degree of bends to be made. Next, multiply the offset measurements by the figure directly under "degree of bend." The above applies to all sizes of conduit.

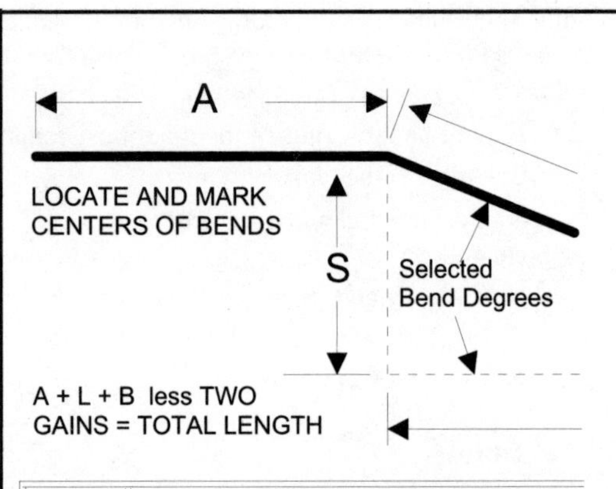

TO FIND UNKNOWN	KNOWN	5-5/8°	11-1/4°	15°	22-1/2°
		TIMES CORRESPONDING TABLE OF MULTIPLIERS FOR			
L	S	10.207	5.126	3.864	2.613
S	L	.098	.195	.259	.383
R	S	10.158	5.027	3.732	2.414
S	R	.098	.199	.268	.414
L	R	1.005	1.02	1.035	1.082
R	L	.995	.981	.966	.933
GAIN PER BEND	RADIUS OF SHOE	.0002	.0006	.0015	.0051

Figure 5-31: Conduit offset bending table.

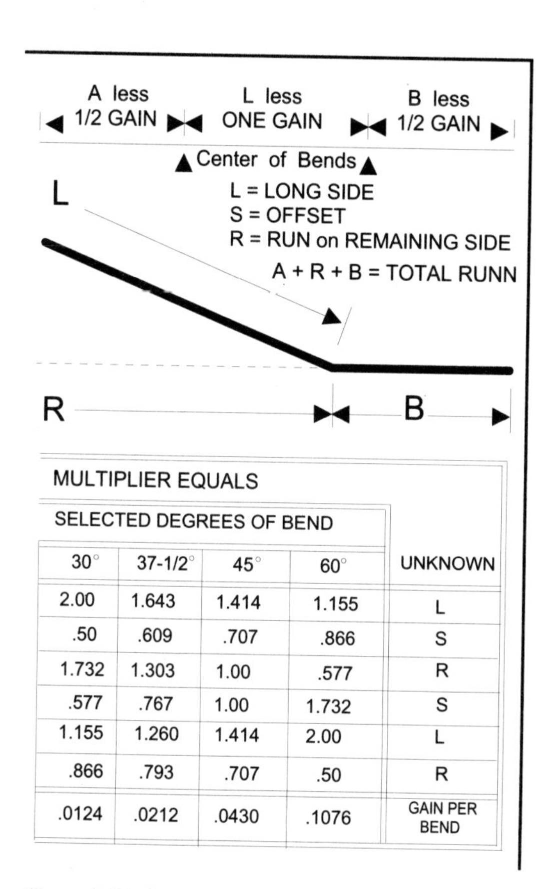

MULTIPLIER EQUALS				
SELECTED DEGREES OF BEND				
30°	37-1/2°	45°	60°	UNKNOWN
2.00	1.643	1.414	1.155	L
.50	.609	.707	.866	S
1.732	1.303	1.00	.577	R
.577	.767	1.00	1.732	S
1.155	1.260	1.414	2.00	L
.866	.793	.707	.50	R
.0124	.0212	.0430	.1076	GAIN PER BEND

Figure 5-31: Conduit offset bending table. ***(Cont.)***

For example, to form an 18-in offset with two 45-degree bends, first make the following calculation to determine the distance between bends.

$$18'' \times 1.414 = 25\tfrac{1}{2} \text{ in}$$

This is the distance between bends and is labeled side "L" (*see* Figure 5-31).

To connect the two ends of an offset to two pieces of conduit already in place, it is necessary to know the overall length (OL) of the offset from end to end, before bending. Note the following equation:

$$(A + L + B) - (2)\,(\text{gains}) = OL$$

The gain is calculated by multiplying the shoe radius by decimal figures shown on the last line under "degrees of bends."

Let's see how an offset might be made in a length of 3-in conduit using 45 degree bends where:

$A = 36$ in

$L = 25\tfrac{1}{2}$ in

$B = 48$ in

$2 \times \text{gain} = 1\tfrac{9}{32}$ in

$OL = (A + L + B - 2)(\text{gains}) = 108\tfrac{7}{32}$ in

Note: This offset uses a 15-inch radius.

$$15 \times .0430 = .645''$$

For two gains, we have:

$$1.290'' = 1\tfrac{9}{32}''$$

Since we already know the long-side dimension of the offset (25 in), to find the amount of offset, refer to the table in Figure 5-31, second line down (to find "S"); move across the row until we come to the 45-degree column. Note the multiplier is .707. Therefore,

$$25 \times .707 = 17.675''$$

This is the amount or height of the offset.

While making the offset, the conduit is positioned in the hydraulic bender as shown in Figure 5-32. In general, the bending shoe makes one 45-degree bend on the first center mark; the conduit is reversed in the bender and the next 45-degree bend is made at the second mark. In doing so, always refer to the bending charts that accompany the bender being used for the operation.

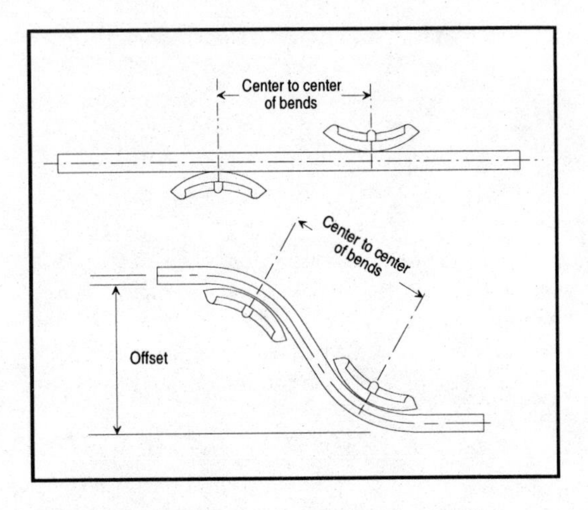

Figure 5-32: Position of conduit in bender for making offsets.

Saddle Bends

As mentioned previously, saddle bends are used to cross a small obstruction or other runs of conduit. Saddle bends in conduit up to 4 in ID may be made in hydraulic conduit benders. In doing so, refer to the charts that accompany the bender. For example, to make a saddle bend on a length of 2-in rigid conduit so that it can pass over a 3-in water pipe with $\frac{1}{4}$-in clearance, three bending operations are required as shown in Figure 5-33.

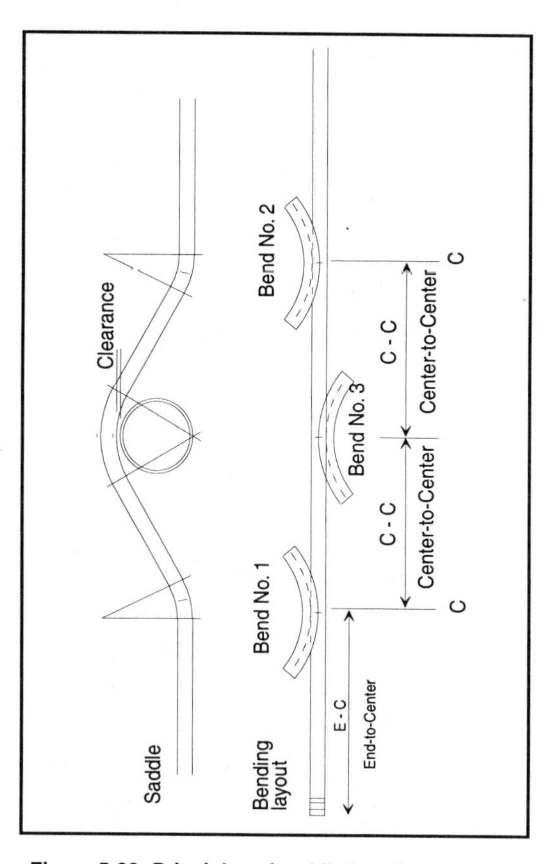

Figure 5-33: Principles of saddle bending.

Straight-Run Conduit	Minimum Length	Bend Spacing	Bend Degrees	Bend Degrees	Bend Degrees
	E-C	C-C	No. 1	No. 2	No. 3
1″	20″	16″	6	6	12
1¼″	20″	16″	7	7	14
1½″	20″	16″	8	8	16
2″	20″	15⅞″	10	10	20
2½″	20″	15¾″	12½	12½	25
3″	20″	15⅝″	15	15	30
3½″	20″	15½″	18	18	36
4″	20″	15½″	20	20	40

Figure 5-34: Saddle bending table.

The bend is calculated by referring to appropriate bending charts. The one shown in Figure 5-34 gives measurements for 2-in conduit with ¼-in clearance. Therefore, look in the left column under "straight-run conduit" and go down the column until the "3-in" row is reached. Moving to the right in this row, it can be seen that the spacing between bends is 15½ in on center, and bends No. 1 and No. 2 will be 15°, while the third bend is 30°.

Using the information found in the bending table, mark the conduit accordingly. Insert the conduit in the bender and make bends No. 1 and No. 2 first — both at 15°. Back off the pump pressure, reverse the conduit in the bender, and then make the third bend to 30°. Release the pump pressure, remove the conduit from the bender and check the saddle for accuracy.

Tricks of the Trade

There are many "tricks of the trade" that you will learn during your career as an electrician. Some of these will be handed down to you by experienced workers; others will be learned through experience. In fact, many electrical workers and most contractors are constantly seeking new methods to improve their efficiency and to make the work go smoother and faster, without sacrificing workmanship.

One of the handiest personal tools used during conduit bending is the small, magnetic torpedo level. This tool should be in the tool pouch of every electrician. In many cases, it can make the difference between a good job and a poor job; that is, obtaining level and plumb conduit runs or not.

Workers have also designed their own custom tools to help in conduit installations. One is taking a pair of vice-grip pliers, welding a small flat piece of iron on the top of the jaws for use with the magnetic level. During multiple bends, the vice grips are positioned at the desired point on the end of a piece of conduit, locked in placed, and then the magnetic level is placed on the flat plate so the bubble may be watched during the bend.

There are also some new commercial tools that are designed for leveling conduit bends.

PVC Conduit Installations

Rigid nonmetallic conduit and fittings (PVC electric conduit) may be used where the potential is 600 V or less in direct earth burial; in walls, floors, ceilings of buildings; in cinder fill, and in damp and dry locations except in certain hazardous locations, for support of fixtures or other equipment, and where subject to physical damage.

PVC conduit can be cut easily at the job site without special tools, although PVC cutters help in cutting square ends. Sizes ½ in through 1½ in can be cut with a fine-tooth saw. For sizes 2 through 6 in, a miter box or similar saw guide should be used to keep the conduit steady and assure a square cut. To assure satisfactory joining, care should be taken not to distort the end of the conduit when cutting.

After cutting, deburr the pipe ends and wipe clean of dust, dirt, and plastic shavings. Deburring is accomplished easily with a pocket knife or file.

One of the important advantages of PVC conduit, in comparison with other rigid conduit materials, is the ease and speed with which solvent cemented joints can be made. The following steps are required for a proper joint:

- Conduit should be wiped clean and dry.
- Apply a full even coat of PVC cement to the end of the conduit. Cement should cover the area that will be inserted in the socket.
- Push conduit and fitting firmly together with slight twisting action until it bottoms and then rotate the conduit in the fitting (about a half turn) to distribute the cement evenly. Avoid cement build-up in ID (inside diameter) of conduit. The cementing and joining operation should not exceed more than 20 seconds. Let dry for approximately 10 minutes.

When the proper amount of cement has been applied, a bead of cement will form at the joint. Wipe the joint with a brush to remove any excess cement. The joints should not be disturbed for ten minutes at room temperatures.

Bending PVC Conduit

Most manufacturers of PVC conduit offer various radius bends in a number of segments. Where special bends are required, PVC conduits are easy to form on the job. Stub-ups, saddles, concentric bends, offsets, and kicks, are all possible with PVC conduit — the same as with metallic conduit.

PVC conduit is bent with the aid of a heating unit. The PVC must be heated evenly over the entire length of the curve. Heating units are available from various sources that are designed specifically for the purpose in sizes to accommodate all conduit diameters. While some heaters use gas for the heat source, most employ infrared heat energy which is most quickly absorbed in the conduit. Small sizes are ready to bend after a few seconds in the "hotbox." Larger diameters require two or three minutes depending on the conditions. Other methods of heating PVC conduit for bending include heating blankets and hot-air blowers. The use of torches or other flame-type devices is not

WARNING!

When working with solvents and glues, provide adequate ventilation to carry off fumes; avoid contact with the skin, and always wear gloves when working with heat.

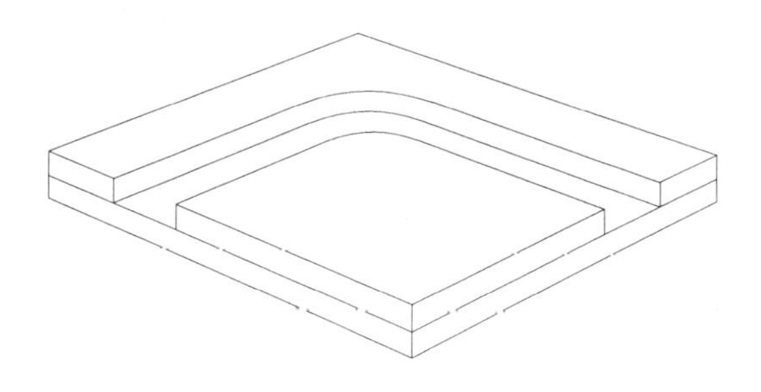

Figure 5-35: A jig can be made by cutting and fitting pieces of plywood to match the desired bend.

recommended. PVC conduit exposed to excessively high temperatures may take on a brownish color. Sections showing evidence of such scorching should be discarded.

If a number of identical bends are required, a jig can be helpful. (*See* Figure 5-35). A simple jig can be made by sawing a sheet of plywood to match the desired bend. Nail to a second sheet of plywood. The heated conduit section is placed in the jig, sponged with water to cool, and it's ready to install. Care should be taken to fully maintain ID of the conduit when handling.

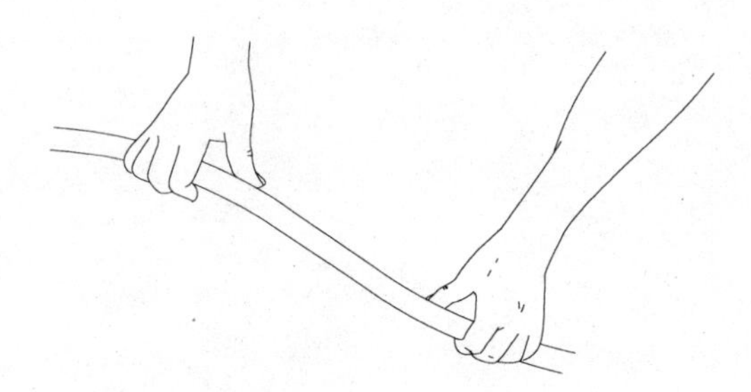

Figure 5-36: Small sizes of PVC conduit may be formed with the hands.

If only a few bends are needed, scribe a chalk line on the floor or workbench. Then match the heated conduit to the chalk line and cool. The conduit must be held in the desired position until relatively cool since the PVC material will tend to go back to its original shape. Templates are also available for any desired bend that complies with the *NEC* requirements.

Another method is to take the heated conduit section to the point of installation and form it to fit the actual installation with the hands. (*See* Figure 5-36). Then wipe a wet rag over the bend (Figure 5-37) to cool it. This method is especially effective in making "blind" bends or compound bends.

Bends in small-diameter PVC conduits (½-1½ in) require no filling for code-approved radii. When bending PVC of 2 in or larger diameter, there is a risk

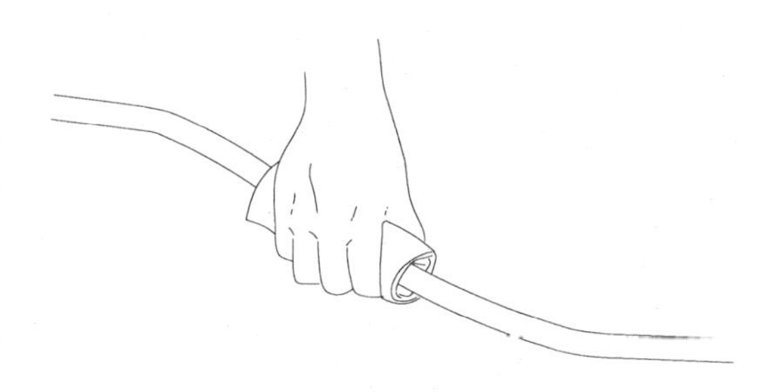

Figure 5-37: After the bend is formed, wipe a wet rag over the bend to cool it.

of wrinkling or flattening the bend. To help eliminate this problem, a plug set is used. A plug is inserted in each end of the piece of PVC being bent; then a hand pump may be used to pressurize the conduit before bending it. The pressure is about 3 to 5 pounds per square inch.

Place airtight plugs in each end of the conduit section before heating. The retained air will expand during the heating process and hold the conduit open during the bending. Do not remove the plugs until the conduit has cooled.

In applications where the conduit installation is subject to constantly changing temperatures and the runs are long, precautions should be taken to allow for expansion and contraction of PVC conduit.

When expansion and contraction are factors, an O-ring expansion coupling should be installed near the fixed end of the run, or fixture, to take up any expansion or contraction that may occur. Confirm the expansion and contraction length available in these fittings as it may vary by manufacturer. Charts are available that indicate what expansion can be expected at various temperature levels. The coefficient of linear expansion of PVC conduit is 0.0034 inch/10ft/°F.

Expansion couplings are seldom required in underground or slab applications. Expansion and contraction may generally be controlled by bowing the conduit slightly or immediate burial. After the conduit is buried, expansion and contraction ceases to be a factor. Care should be taken, however, in constructing a buried installation. If the conduit should be left exposed for an extended period of time during widely variable temperature conditions, allowance should be made for expansion and contraction.

In above-ground installations, care should be taken to provide proper support of PVC conduit due to its semirigidity. This is particularly important at high temperatures. Distance between supports should be based on temperatures encountered at the specific installation. Charts are available that clearly outline at what intervals support is required for PVC conduit at various temperature levels.

Cutting, Reaming, and Threading Conduit

Rigid conduit, intermediate metal conduit (IMC), and electrical metallic tubing (EMT) are manufactured in standard 10-ft lengths. When installing conduit, it is cut to fit the job requirements.

In general, there are three methods used to cut metal conduit:

- Hacksaw method
- Pipe-cutter method
- Power saw method

To cut conduit with a hacksaw, use the following steps:

Step 1. Inspect the blade of the hacksaw and replace it if needed. A blade with 18, 24, or 32 teeth per in is recommended for conduit. Use 24 or 32 teeth per in for EMT and 18 teeth per in for rigid conduit and IMC. The teeth of the blade should be pointed toward the front of the saw.

Step 2. Secure the conduit in a pipe vise.

Step 3. Rest the middle of the hacksaw blade on the conduit where the cut is to be made. Position the saw so the end of the blade is pointing slightly downward and the handle is slighly up. Push forward gently until the cut is started. Make even strokes until the cut is finished.

Step 4. Check the cut. The end of the conduit should be straight and smooth.

A pipe cutter can also be used to cut conduit. Use the following steps when operating a pipe cutter:

Step 1. Secure the conduit in a pipe vise and mark a place for the cut.

Step 2. Open the cutter and place it over the conduit with the cutter wheel on the mark.

Step 3. Tighten the cutter by rotating the screw handle, but do not overtighten the cutter.

Step 4. Rotate the cutter counterclockwise to start the cut as shown in Figure 5-38.

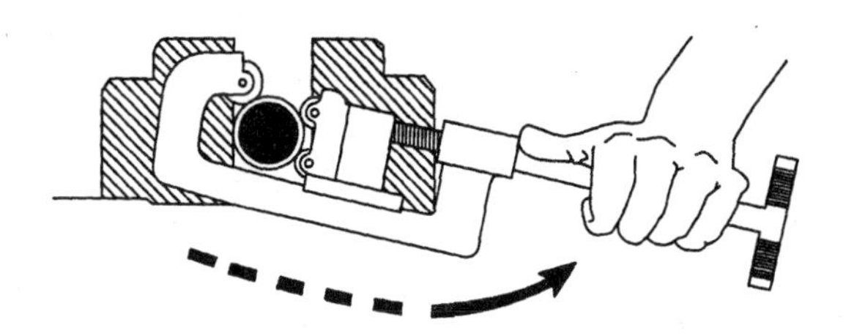

Figure 5-38: Rotating the pipe cutter for the starting cut.

Step 5. Tighten the cutter handle one-quarter turn for each full turn around the conduit. Again, make sure that you don't overtighten.

Step 6. Add a few drops of cutting oil to the groove and continue cutting until finished.

Step 7. When the cut is almost finished, stop cutting and snap the conduit to finish the cut. This prevents a ridge from forming on the inside of the conduit.

Step 8. Clean the conduit and cutter with a shop towel or rag.

Reaming Conduit

After the conduit is cut, the inside edge is sharp and usually contains burrs. This edge will damage the conductor insulation and must be removed to avoid this damage. A reamer is used for this operation and two types in common use are shown in Figure 5-39. One type has a ratchet handle and the other has a shank for insertion into a bit brace.

Use the following steps to ream the inside edge of a length of conduit:

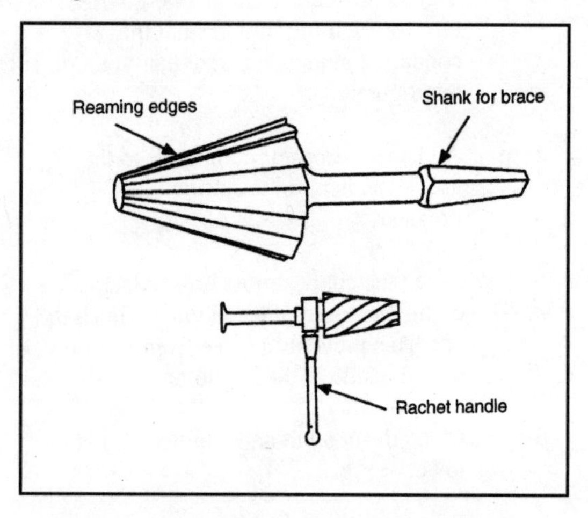

Figure 5-39: Two types of conduit reamers.

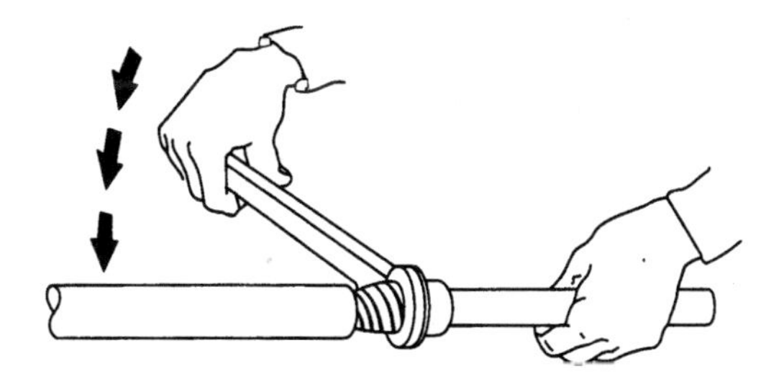

Figure 5-40: Rotation direction for reaming.

Step 1. Place the conduit in a pipe vise.

Step 2. Insert the reamer tip at the end of the conduit.

Step 3. Apply light forward pressure and start rotating the reamer as shown in Figure 5-40. Note that the reamer can be damaged if it is rotated in the wrong direction. The reamer should "bite" as soon as the proper pressure is applied.

Step 4. Remove the reamer by pulling back on it while continuing to rotate it in the same direction for reaming. In other words, don't reverse the rotation.

Step 5. Check the progress and then reinsert the reamer if necessary. Rotate the reamer until the inside edge is smooth. You should stop when all burrs have been removed.

If a conduit reamer is not available, a half-round metal-cutting file can be used for rigid conduit and IMC. The smaller sizes of EMT may be reamed with the nose of side-cutting pliers, small hand reamers, or even the shank of a square-shanked screwdriver.

Joining Conduit

After conduit is cut and reamed, it must be properly joined and secured in place before conductors are pulled in.

EMT is joined by couplings. Two types are in common use:

- Set-screw couplings
- Compression couplings

Both types are shown in Figure 5-41 and a brief description of each follows.

As its name implies, the set-screw coupling relies on set screws to hold the EMT to the coupling. This type of coupling does not provide a seal and is not

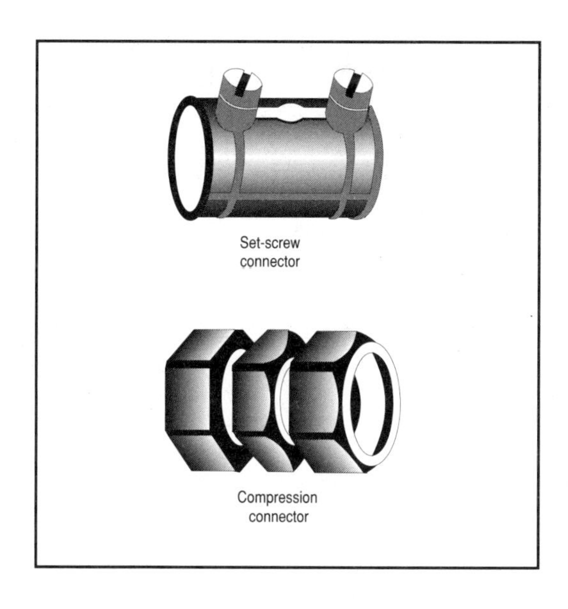

Figure 5-41: Two types of EMT couplings.

permitted to be used in wet locations. However, this type of coupling can be imbedded in concrete.

Compression couplings provide a tight seal around the conduit, and may be used in some wet locations as stated in the *NEC*.

Galvanized rigid steel conduit (GRC) threadless connectors can also be used with the larger sizes of

EMT; that is, $2\frac{1}{2}$ in and larger, because in these sizes, the outside diameter matches the outside diameter of rigid steel conduit. However, the reverse is not permitted; that is, EMT connectors may not be used to connect rigid steel conduit.

Although threadless couplings and connectors may be used under certain conditions with rigid steel conduit, there are many cases when rigid steel conduit and IMC must be threaded.

The tool used to cut conduit threads is called a "thread-cutting die" or just plain "die." Conduit dies are designed to cut threads with a taper of $\frac{3}{4}$ in per foot. The number of threads per inch varies from 8 to 18, depending upon the diameter of the conduit. A thread gauge is used to measure the number of threads per inch.

The threading dies are contained in a die head. The die head can be used with a hand-operated ratchet threader (Figure 5-42) or with a portable power drive.

To thread conduit with a hand-operated threader, perform the following steps:

Step 1. Insert the conduit in a pipe vise. Make sure the vise is fastened to a strong surface. Place supports, if necessary, to help secure the conduit.

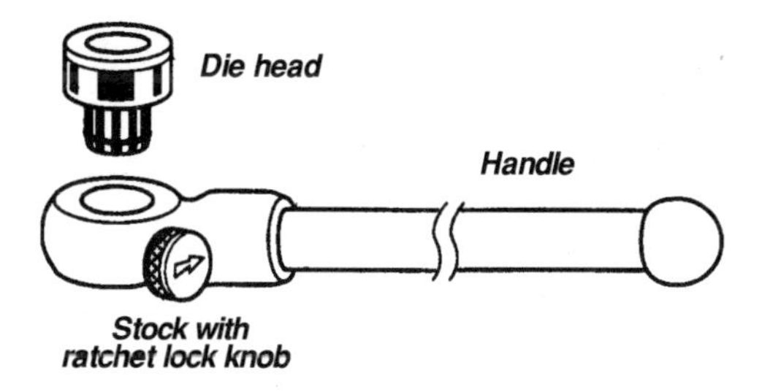

Figure 5-42: Hand-operated ratchet conduit threader.

broken teeth. Never use a damaged die.

Step 3. Insert the die securely in the head. Make sure the proper die is in the appropriately numbered slot on the head.

Step 4. Determine the correct thread length to cut for the conduit size used.

Step 5. Lubricate the die with cutting oil at the beginning and throughout the threading operation.

Step 6. Cut threads to the proper length. Make sure that the conduit enters the tapered side of the die. Apply pressure and start turning the head. You should back off the head each quarter-turn to clear away chips.

Step 7. Remove the cutter when the cut is complete. Threads should be cut only to the length of the die. Overcutting will leave the threads exposed to the elements and will corrode.

Step 8. Inspect the threads to make sure they are clean, sharp, and properly made. Use a thread gauge to measure the threads. The finished end should allow for a wrench-tight fit with not more than one thread exposed.

Step 9. Ream the conduit once again to remove any burrs and edges. Cutting oil must be removed from the inside and outside of the conduit.

Die heads can also be used with portable power drives; follow the same steps as listed previously.

Threading machines are often used on larger sizes of conduit and where much threading is to be done. Threading machines hold and rotate the conduit while the die is fed onto the conduit for cutting. When using a conduit threading machine, make sure the threader's legs are secured properly and follow the manufacturer's instructions.

Summary

Many conduit installations are visible; that is, run exposed for everyone to see. Consequently, electricians must take special care to ensure that all exposed conduit runs are parallel, level, and plumb. Nothing else will do! This is one phase of the electrical construction industry where electricians have an opportunity to "show their stuff." In fact, an expert installation of a conduit system is not unlike a work of art.

Learn the basics of conduit bending and installations, put your knowledge to practical use, and take pride in your abilities to perform a conduit installation that takes second place to none. Of course, contractors and clients want speed, but if you have a good basic knowledge of conduit bending and you then put this knowledge to use, your craftmanship will let you bend conduit smarter and faster — giving a good-looking installation along with speed to satisfy your employer, the building owners, and all concerned.

Chapter 6

OUTLET, JUNCTION, AND PULL BOXES

On every job, a great number of boxes is required for outlets, switches, pull and junction boxes. All of these must be sized, installed and supported to meet current *NEC* requirements. Since the *NEC* limits the number of conductors allowed in each outlet or switch box — according to its size — electricians must install boxes large enough to accommodate the number of conductors that must be spliced in the box or fed through. Therefore, a knowledge of the various types of boxes and the volume of each is essential.

A box or fitting must be installed at:

- Each conductor splice point
- Each outlet, switch point, or junction point
- Each pull point for the connection of conduit or other raceways

Furthermore, boxes or other fittings are required when a change is made from conduit to open wiring or cable. Electrical workers also install pull boxes in raceway systems to facilitate the pulling of conductors.

Many boxes used in electrical installations are made from sheet steel with the surface of the metal boxes galvanized to resist corrosion and to provide continuous bonding throughout the system. Nonmetallic boxes made of PVC or Bakelite are also used to some extent. Various types of outlet boxes are shown in Figure 6-1.

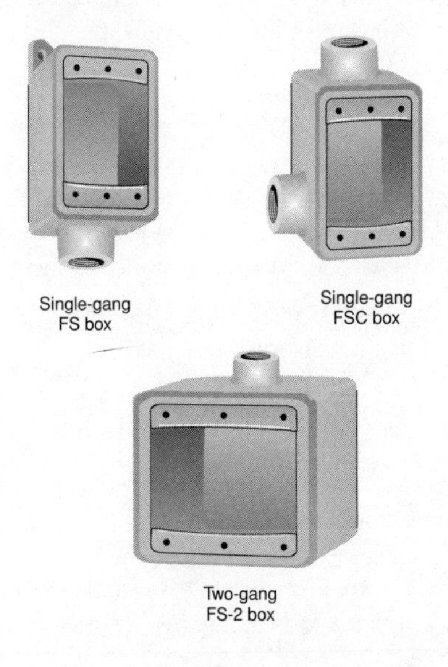

Figure 6-1: Various types of outlet boxes.

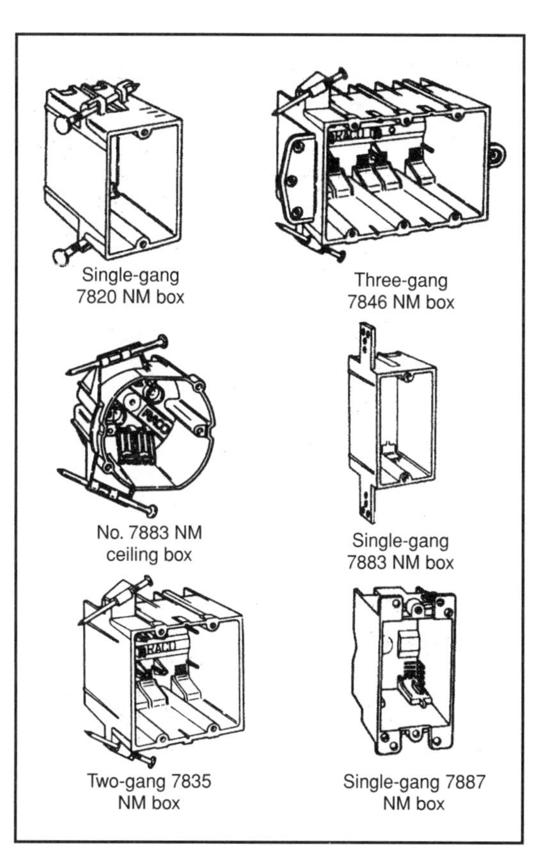

Figure 6-1: Various types of outlet boxes. ***(Cont.)***

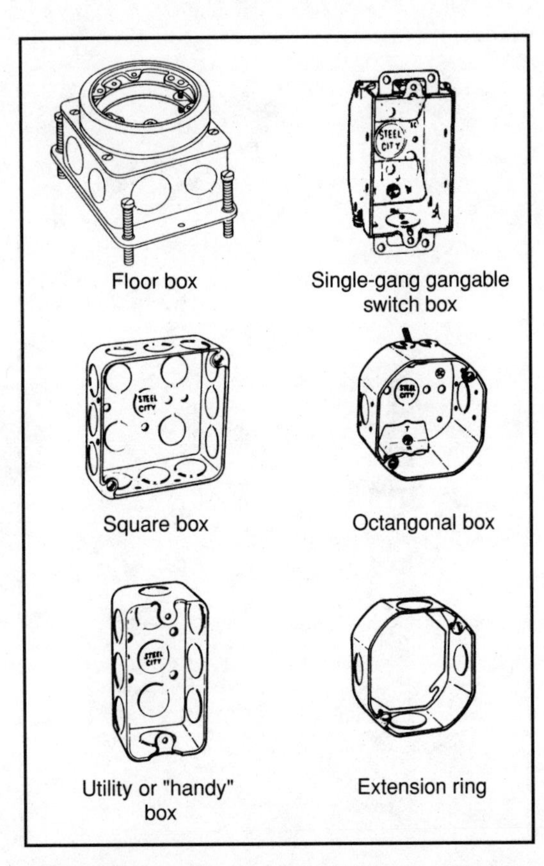

Figure 6-1: Various types of outlet boxes. ***(Cont.)***

Metal outlet boxes are made with removable circular sections called knockouts or pryouts. Knockouts are removed to make openings for conduit or cable connections to the boxes. The basic knockout is a half-way cut disk that is easily removed when sharply hit by a hammer and punch. Some knockouts, however, are concentric, in which case these are several sections which can be removed to fit the desired conduit or connector.

A pryout is a variation of the knockout. In the former, a slot is cut into the center of the metal tab. To remove a pryout, insert a screwdriver blade into the slot and twist it to break the solid tab as shown in Figure 6-2.

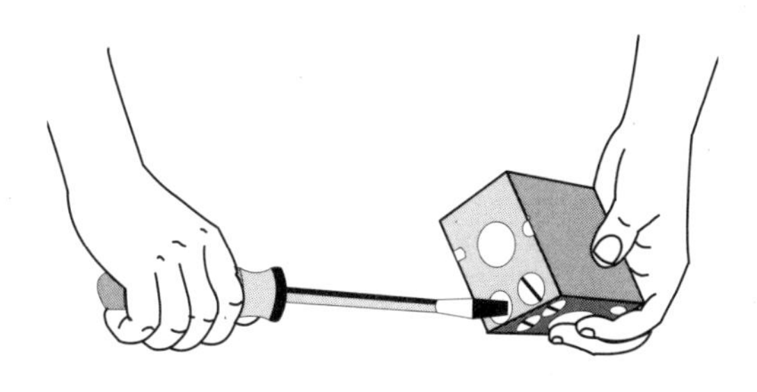

Figure 6-2: Method used to remove a pryout from an outlet box.

In each case — raceways, outlet boxes, pull and junction boxes — the *NEC* specifies specific maximum fill requirements; that is, the area of conductors in relation to the area of the box, fitting, or raceway system.

Sizing Outlet Boxes

In general, the maximum number of conductors permitted in standard outlet boxes is listed in Table 370-16(a) of the *NEC*. These figures apply where no fittings or devices such as fixture studs, cable clamps, switches, or receptacles are contained in the box and where no grounding conductors are part of the wiring within the box. Obviously, in all modern residential wiring systems there will be one or more of these items contained in the outlet box. Therefore, where one or more of the above mentioned items are present, the number of conductors must be one less than shown in the tables. For example, a deduction of two conductors must be made for each strap containing a device such as a switch or duplex receptacle; a further deduction of one conductor must be made for one or more grounded conductors entering the box. A 3-in × 2-in × 2¾-in box for example, is listed in the table as containing a maximum number of six No. 12 wires. If the box contains cable clamps and a duplex receptacle, three wires will have to be deducted from the total of six — providing for only three No. 12 wires. If a ground wire is used, only two No. 12 wires may be used.

Figure 6-3 illustrates one possible wiring configuration for outlet boxes and the maximum number of conductors permitted in them as governed by Section

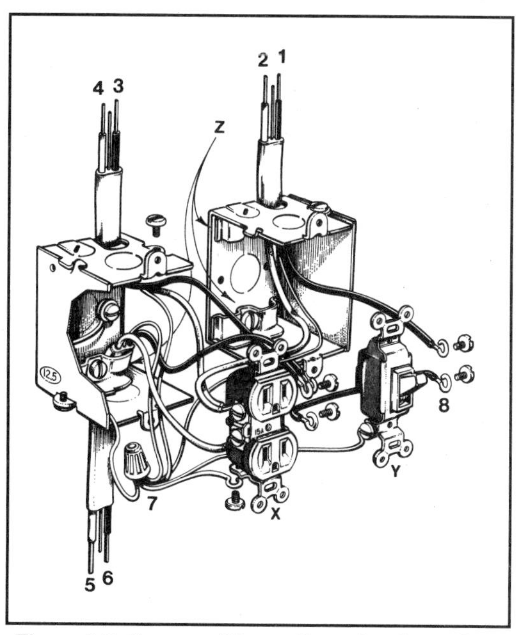

Figure 6-3: One possible configuration for outlet boxes.

370-16 of the *NEC*. This example shows two single-gang switch boxes joined or "ganged" together to hold a single-pole toggle switch and a duplex receptacle. This type of arrangement is likely to be found above a workbench where the duplex receptacle is provided for small power tools and the single-pole switch could be used to control overhead lighting.

Since Table 370-16(a) gives the capacity of one $3 \times 2 \times 2\frac{1}{4}$-in device box as 12.5 in^3, the total capacity of both boxes in Figure 6-3 is 25 in^3. These two boxes have a capacity to allow 10 No. 12 AWG conductors, or 12 No. 14 AWG conductors, less the deductions as listed below.

- Two conductors must be deducted for each strap-mounted device. Since there is one duplex receptacle (X) and one single-pole toggle switch (Y), four conductors must be deducted from the total number stated in the above paragraph.
- Since the combined boxes contain one or more cable clamps (Z), another conductor must be deducted. Note that only one deduction is made for similar clamps, regardless of the number. However, any unused clamps may be removed to facilitate the electrical worker's job; that is, allowing for more work space.

- The equipment grounding conductors, regardless of the number, count as one conductor only.

Therefore, to comply with the *NEC*, and considering the combined deduction of six conductors, only four No. 12 AWG conductors (six No. 14 AWG conductors) may be installed in the outlet-box configuration in Figure 6-3.

Figure 6-3 shows three nonmetallic-sheathed (NM or Romex) cables, designated 12/2 with ground, entering the ganged outlet boxes. This is a total of six current-carrying conductors and three ground wires, for a total of nine. Is this arrangement in violation of the *NEC*? Yes, because the total number of conductors exceeds the *NEC* limits. However, if No. 14 AWG conductors were installed rather than No. 12, the configuration will comply with the 1996 *NEC*. Another alternative is to go to $3 \times 2 \times 3\frac{1}{2}$-in device boxes which would then have a total of 36 in^3 for the two boxes.

Also note the jumper wire in Figure 6-3; this is numbered "8" in the drawing. Conductors that both originate and end in the same outlet box are exempt from being counted against the allowable capacity of an outlet box. This jumper wire (8) taps off one terminal of the duplex receptacle to furnish a "hot wire" to the single-pole toggle switch. Therefore, this wire originates and terminates in the same set of ganged boxes and is not counted against the total number of conductors. By the same token, the three grounding conductors extending from the wire nut to the individual grounding screws on the devices originate and terminate in the same set of boxes. These conductors are also exempt from being counted with the total. Incidentally, the wire nut has a crimp connector beneath; wire nuts alone are not allowed to connect equipment grounding conductors.

A pictorial definition of stipulated conditions as they apply to Section 370-16 of the *NEC* is shown in Figure 6-4. Figure 6-4A illustrates an assortment of raised covers and outlet box extensions. These components, when combined with the appropriate outlet boxes, serve to increase the usable work space. Each type is marked with their cubic-inch capacity which may be added to the figures in *NEC* Table 370-16(a) to calculate the increased number of conductors allowed.

Figure 6-4B shows components that may be used in outlet boxes without affecting the total number of conductors. Such items include grounding clips and screws, wire nuts and cable connectors when the latter is inserted through knockout holes in the outlet box and secured with lockouts. Prewired fixture conductors are not counted against the total number of allowable conductors in an outlet box where there are

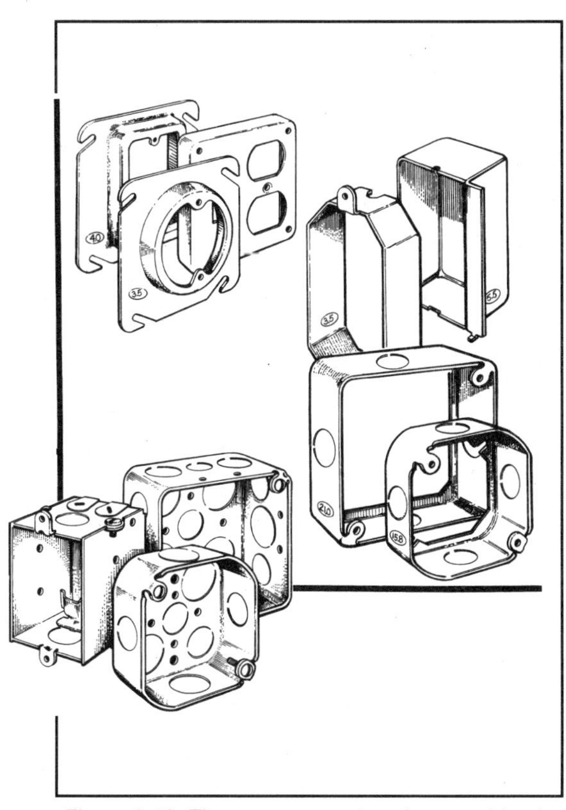

Figure 6-4A: These components, when combined with the appropriate outlet boxes, serve to increase the usable work space within the box.

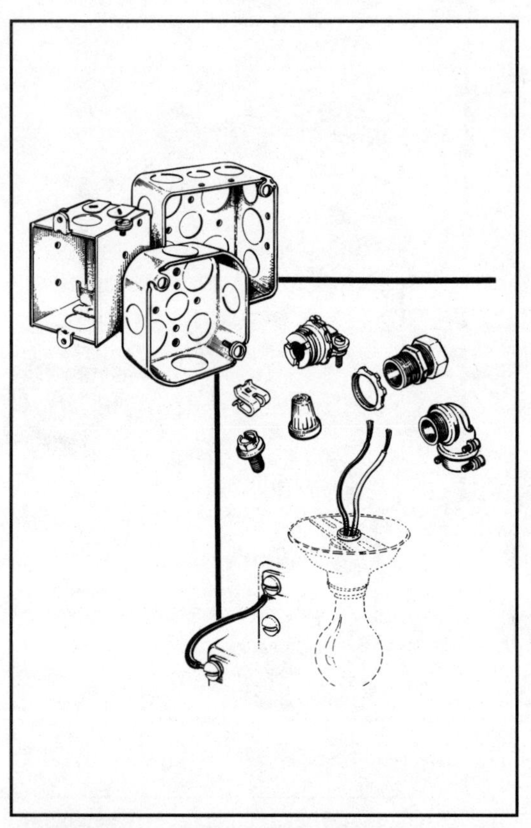

Figure 6-4B: Items that do not affect the capacity of outlet boxes.

not over four wires smaller than No. 14 AWG. Conductors originating and ending in the same box need not be counted.

Figure 6-4C on the next page shows typical wiring configurations that must be counted as conductors when calculating the total capacity of outlet boxes. Further details of these configurations are as follows:

- Each wire passing through a box without a splice or tap is counted as one conductor. Therefore, a cable containing two wires (shown in the drawing) that passes in and out of an outlet box without a splice or tap is counted as two conductors.
- Wires that enter and terminate in the same box are charged as individual conductors and in this case, the total charge would be two conductors.
- Each wire that enters a box and is either spliced or connected to a terminal, and then exits again, is counted as two conductors. In the case of two 2-wire cables, the total conductors charged will be four.
- When one or more grounding wires enters the box and are joined, a deduction of only one conductor is required, regardless of their number.

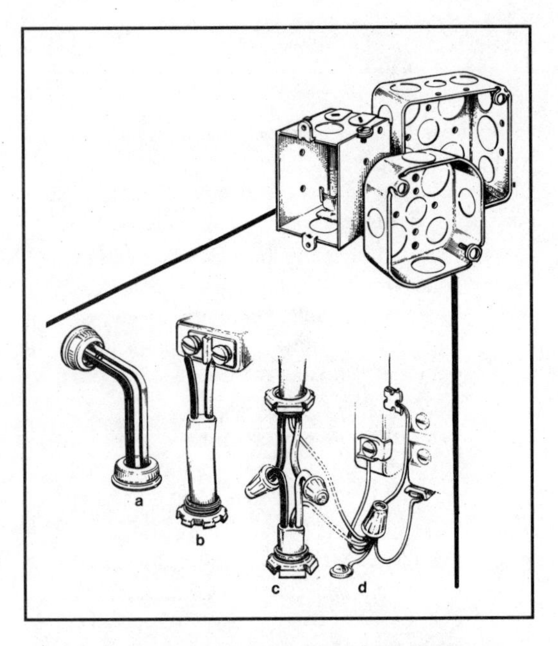

Figure 6-4C: Configurations that must be counted as conductors when calculating box capacity.

Two conductors must be deducted for each strap-mounted device, like duplex receptacles and wall switches as shown in Figure 6-4D.

Figure 6-4E shows further components that require deduction adjustments from those specified in

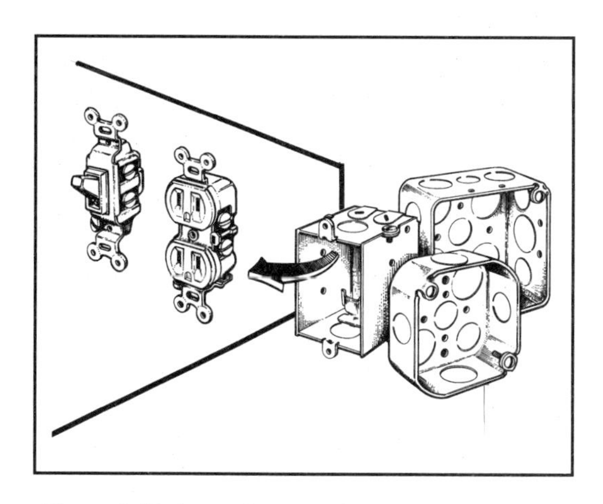

Figure 6-4D: Items that must be counted as 2 conductors when sizing box capacity.

Table 370-16(a). Such items include fixture studs, hickeys, and fixture-stud extensions and one conductor must also be deducted from the total for each type of fitting used. A deduction of one conductor is made when one or more internally-mounted cable clamps are used.

To better understand how outlet boxes are sized, let's take two No. 12 AWG conductors installed in ½-in EMT and terminating into a metallic outlet box containing one duplex receptacle. What size of outlet box will meet NEC requirements?

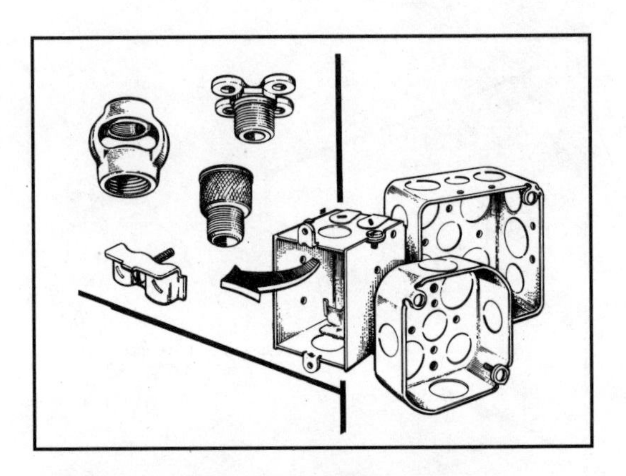

Figure 6-4E: Items that must be counted as 1 conductor when sizing outlet-box capacity.

The first step is to count the total number of conductors and equivalents that will be used in the box — following the requirements specified in *NEC* Section 370-16.

Step 1. Calculate the total number of conductors and equivalents.

One receptacle	= 2 conductors
Two #12 conductors	= 2 conductors
Total #12 conductors	= 4

Step 2. Determine amount of space required for each conductor.

NEC Table 370-16(b) gives the box volume required for each conductor:

$$\text{No. 12 AWG} = 2.25 \text{ in}^3$$

Step 3. Calculate the outlet-box space required by multiplying the number of in^3 required for each conductor by the number of conductors found in No. 1 above.

$$4 \times 2.25 = 9.00 \text{ in}^3$$

Once you have determined the required box capacity, again refer to *NEC* Table 370-16(a) and note that a $3 \times 2 \times 2\frac{1}{4}$-in box comes closest to our requirements. This box size is rated for 10.5 in^3.

Where four No. 12 conductors enter the box, two additional No. 12 conductors must be added to our previous count for a total of (4 + 2 =) 6 conductors.

$$6 \times 2.25 = 13.5 \text{ in}^3$$

Again, refer to *NEC* Table 370-16(a) and note that a $3 \times 2 \times 3\frac{3}{4}$-in device box, with a rated capacity of 14.0 in^3, is the closest device box that meets *NEC* requirements. Of course, any box with a larger capacity is permitted.

CONDUIT BODIES, PULL BOXES AND JUNCTION BOXES

The *NEC* specifically states that at each splice point, or pull point for the connection of conduit or other raceways, a box or fitting must be installed. The *NEC* specifically considers conduit bodies, pull boxes, and junction boxes and specifies the installation requirements as listed in the table in Figure 6-5.

Conduit bodies provide access to the wiring through removable covers. Typical examples are Types T, C, X, L, and LB. Conduit bodies enclosing No. 6 or smaller conductors must have an area twice that of the largest conduit to which they are attached, but the number of conductors within the body must not exceed that allowed in the conduit. If a conduit body has entry for three or more conduits such as Type T or X, splices may be made within the conduit body. Splices may not be made in conduit bodies having one or two entries unless the volume is sufficient to qualify the conduit body as a junction box or device box.

When conduit bodies or boxes are used as junction boxes or as pull boxes, a minimum size box is required to allow conductors to be installed without

CONDUIT BODIES, PULL AND JUNCTION BOXES		
Application	**Requirements**	**NEC Reference**
Accessible	Must provide access to wiring inside of box without removing any part of the building.	Section 370-29
Cellular floor raceways	Must be leveled to the floor grade and sealed to prevent free entry of water or concrete.	Section 354-13
Conduit bodies, size of	Those enclosing No. 6 or smaller conductors must have a cross-sectional area not less than twice that of the raceway system to which the conduit bodies are attached.	Section 370-16(c)

Figure 6-5: Summary of NEC installation requirements for conduit bodies, pull and junction

CONDUIT BODIES, PULL AND JUNCTION BOXES *(Cont.)*		
Application	**Requirements**	**NEC Reference**
Conduit bodies, maximum number of conductors	Must not exceed the allowable fill for the attached conduit.	Section 370-16(c)
Junction and pull boxes	Must comply with (a) through (d) of this *NEC* Section.	Section 370-28
Junction and pull boxes over 6 ft	Conductors must be cabled or racked up.	Section 370-28(b)
Over 600 V	Special requirements apply to boxes used on systems of over 600 V.	Article 370 Part D
Size, straight pulls	Not less than 8 times the trade diameter of the largest raceway.	Section 370-28(a)(1)
Size, angle or U pulls	Not less than 6 times the trade diameter of the largest raceway.	Section 370-28(a)(2)

Figure 6-5: Summary of NEC installation requirements for conduit bodies, pull and junction

undue bending. The calculated dimensions of the box depend on the type of conduit arrangement and on the size of the conduits involved.

Sizing Pull And Junction Boxes

Figure 6-6 shows a junction box with several conduits entering it. Since 4-in conduit is the largest size in the group, the minimum length required for the box can be determined by the following calculation:

Trade size of conduit × 8 (as per *NEC*) = minimum length of box

$$4'' \times 8 = 32''$$

Therefore, this particular pull box must be at least 32 in long. The width of the box, however, need be only of sufficient size to enable locknuts and bushings to be installed on all the conduits or connectors entering the enclosure.

Junction or pull boxes in which the conductors are pulled at an angle, as shown in Figure 6-7 on page 215, must have a distance of not less than six times the trade diameter of the largest conduit. The distance must be increased for additional conduit entries by the amount of the sum of the diameter of all other conduits entering the box on the same side, that is, the wall of the box. The distance between raceway entries enclosing the same conductors must not be less

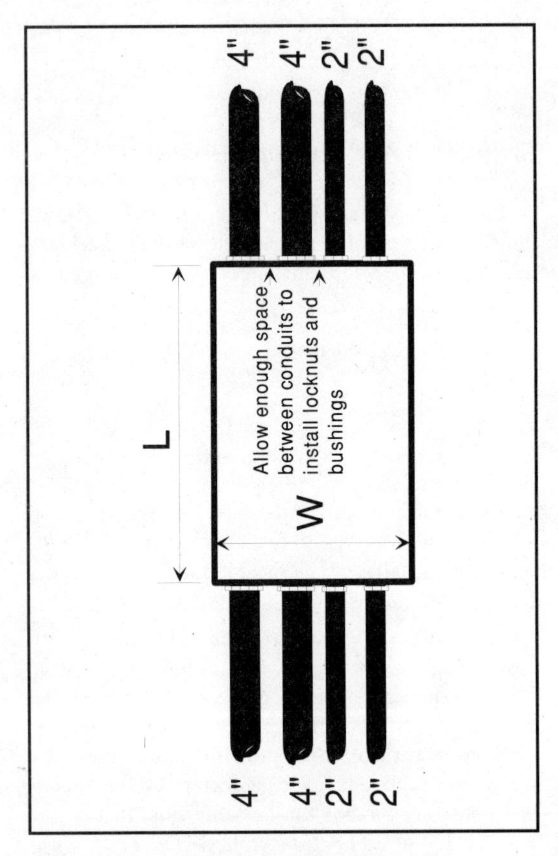

Figure 6-6: Pull box used on straight pulls.

than six times the trade diameter of the largest conduit.

Since the 4-in conduit is the largest of the lot in this case,

$$L_1 = 6 \times 4 + (3 + 2) = 29''$$

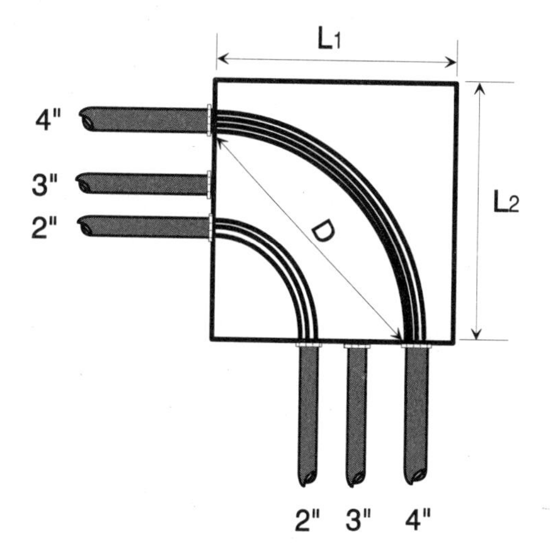

Figure 6-7: Pull box with conduit runs entering and leaving at right angles.

Since the same number and sizes of conduit is located on the adjacent wall of the box, L_2 is calculated in the same way; therefore, $L_2 = 29$ in.

The distance (D) = 6 × 4 or 24 inches and this is the minimum distance permitted between conduit entries enclosing the same conductor.

The depth of the box need only be of sufficient size to permit locknuts and bushings to be properly installed. In this case, a 6-in deep box would suffice.

If the conductors are smaller than No. 4, the length restriction does not apply.

Figure 6-8 shows another straight-pull box. What is the minimum length if the box has one 3-in conduit and two 2-in conduits entering and leaving the box? Again, refer to *NEC* Section 370-28(a)(1) and find that the minimum length is 8 times the largest conduit size which in this case is:

$$8 \times 3 \text{ in} = 24 \text{ in}$$

Let's review the installation requirements for pull or junction boxes with angular or U pulls. Two conditions must be met in order to determine the length and width of the required box.

The minimum distance to the opposite side of the box from any conduit entry must be at least six times the trade diameter of the largest raceway.

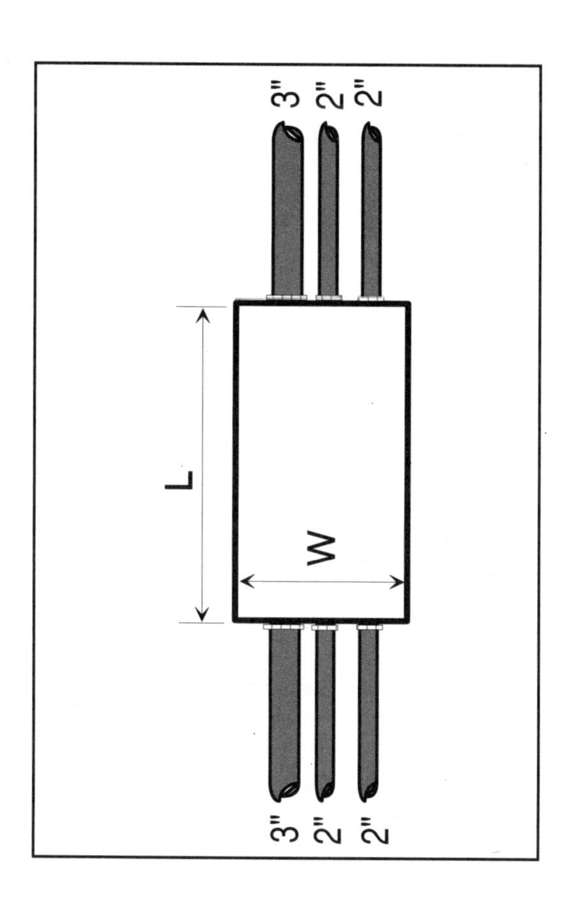
3"
2"
2"
L
W
3"
2"
2"

The sum of the diameters of the raceways on the same wall must be added to this figure.

Figure 6-9 shows the minimum length of a box with two 3-in conduits, two 2-in conduits, and two 1½-in conduits in a right-angle pull. The minimum length based on this configuration is:

6 × 3 in	=	18 in
1 × 3 in	=	3 in
2 × 2 in	=	4 in
2 × 1½ in	=	3 in
		28 in

Since the number and size of conduits on the two sides of the box are equal, the box is square and has a minimum size dimension of 28 in. However, the distance between conduit entries must now be checked to ensure that *all NEC* requirements are met; that is, the spacing (D) between conduits enclosing the same conductor must not be less than six times the conduit diameter. Again refer to Figure 6-9 and note that the 1 ½-in conduits are the closest to the left-hand corner of the box. Therefore, the distance (D) between conduit entries must be:

$$6 \times 1\tfrac{1}{2}\ \textit{inches} = 9\ \text{in}$$

The next group is the two 2-in conduits which is calculated in a similar fashion; that is:

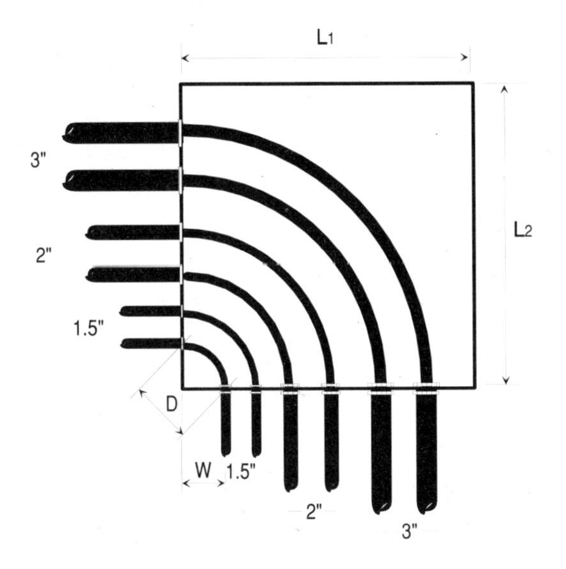

Figure 6-8: Typical straight-pull box.

$$6 \times 2 = 12 \text{ in}$$

The remaining raceways in this example are the two 3-in conduits and the minimum distance between the 3-in conduit entries must be:

$$6 \times 3 = 18 \text{ in}$$

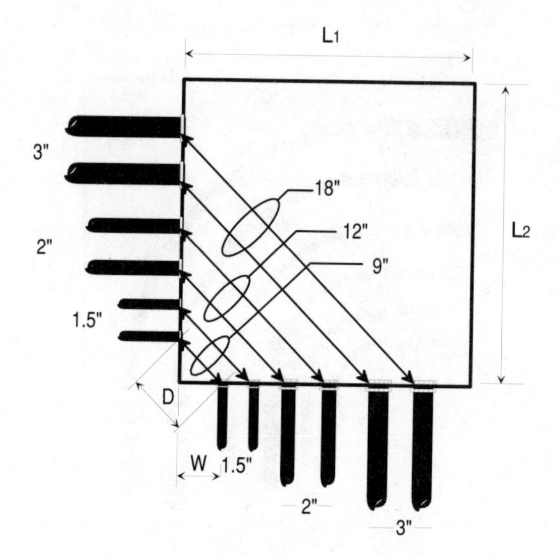

Figure 6-9: Minimum size pull box for angle conduit entries.

A summary of the conduit-entry distances is presented in Figure 6-10. However, some additional math is required to obtain the spacing (w) between the conduit entries. For example, the distance from the corner of the pull box to the center of the conduits (w) may be found by the following equation:

$$\text{Spacing} = \frac{\textit{Diagonal distance}\,(D)}{\sqrt{2}}$$

Consequently, the spacing (w) for the 1½-in conduit may be determined using the following equation:

$$\frac{9}{\sqrt{2}} = \frac{9}{1.414} = 6.4 \text{ in}$$

Therefore, the spacing (w) is 6.4 in. This distance is measured from the left lower corner of the box in each direction — both vertically and horizontally — to obtain the center of the first set of 1½-in conduits. This distance must be added to the spacing of the other conduits including locknuts or bushings.

A rule-of-thumb is to allow ½ in clearance between locknuts.

Using all information calculated thus far, and using Figure 6-10 as reference, the required measurements of the pull box may be further calculated as follows:

Step 1. Calculate space (w):

$$D = 6 \times 1\tfrac{1}{2}\text{-in} = 9 \text{ in}$$

Step 2. Divide this number (9 in) by the square root of 2 (1.414) and make the following calculation:

$$w = \frac{9\ inches}{1.414} = 6.4 \text{ inches}$$

Step 3. Measure from the left lower corner of the pull box over 6.4 in to obtain the center of the knockout for the first $1\frac{1}{2}$-in conduit. Measure up (from the lower left corner) to obtain the center of the knockout for this same cable run on the left side of the pull box.

Step 4. Since there are two 3-in (inside diameter) conduits, each with an outside diameter of approximately 4.25 in, the space for these two conduits can be found by the following equation:

$$2 \times 4.25 = 8.5 \text{ in}$$

Step 5. The space required for the two 2-in (inside diameter) conduits, each with an outside diameter of approximately 3.12 in, may be determined in a similar manner; that is:

$$2 \times 3.12 = 6.24 \text{ in}$$

Step 6. The space required for the two 1.5-in (inside diameter) conduits, each with an outside diameter of approximately 2.62 in, may be determined using the same equation:

$$2 \times 2.62 = 5.24 \text{ in}$$

Step 7. To find the required space for locknuts and bushings, multiply 0.5 in by the total number of conduit entries on one side of the box. Since there are a total of 6 conduit entries, use the following equation:

$$6 \times .5 = 3.0 \text{ in}$$

Step 8. Add all figures obtained in Steps 2 through 7 together to obtain the total required length of the pull box.

Clear space (w)	=	6.4 in
1.5-inch conduits	=	5.24 in
2-inch conduits	=	6.24 in
3-inch conduits	=	8.5 in
Space between locknuts	=	3.0 in
Total length of box	=	29.38 in

Since the same number and size of conduits enter on the bottom side of the pull box and leave, at a right angle, on the left side of the pull box, the box will be square. Furthermore, although a box exactly 29.38 in will suffice for this application, the next larger standard size is 30 in; this should be the size pull box selected. Even if a "custom" pull box is made in a sheet-metal shop, the workers will still probably make it an even 30 in unless specifically ordered otherwise.

Cabinets And Cutout Boxes

NEC Article 373 deals with the installation requirements for cabinets, cutout boxes, and meter sockets. In general, where cables are used, each cable must be secured to the cabinet or cutout box by an approved method. Furthermore, the cabinets or cutout boxes must have sufficient space to accommodate all conductors installed in them without crowding.

NEC Table 373-6(a) gives the minimum wire-bending space at terminals along with the width of wiring gutter in inches. A summary of *NEC* requirements for the installation of cabinets and cutout boxes follows:

- Table 373-6(a) must apply where the conductor does not enter or leave the enclosure through the wall opposite its terminal.

- *Exception No. 1 states:* A conductor must be permitted to enter or leave an enclosure through the wall opposite its terminal provided the conductor enters or leaves the enclosure where the gutter joins an adjacent gutter that has a width that conforms to Table 373-6(b) for that conductor.
- *Exception No. 2 states:* A conductor not larger than 350 kcmil must be permitted to enter or leave an enclosure containing only a meter socket(s) through the wall opposite its terminal, provided the terminal is a lay-in type where either: (a) The terminal is directly facing the enclosure wall and offset is not greater than 50 percent of the bending space specified in Table 373-6(a), or (b) The terminal is directed toward the opening in the enclosure and is within a 45-degree angle directly facing the enclosure wall.
- Table 373-6(b) must apply where the conductor enters or leaves the enclosure through the wall opposite its terminal.

NEC Article 374 covers the installation requirements for auxiliary gutters, which are permitted to supplement wiring spaces at meter centers, distribution centers, and similar points of wiring systems and may enclose conductors or busbars but must not be used to enclose switches, overcurrent devices, appliances, or other similar equipment.

In general, auxiliary gutters must not contain more than 30 current-carrying conductors at any cross section. The sum of the cross-sectional areas of all contained conductors at any cross section of an auxiliary gutter must not exceed 20 percent of the interior cross-sectional area of the auxiliary gutter. Conductors installed in conduits and tubing must not exceed 40 percent fill. Auxiliary gutters are limited to only 20 percent.

When dealing with auxiliary gutters, always remember the number "30." This is the maximum number of conductors allowed in any auxiliary gutter regardless of the cross-sectional area. This question will be found on almost every electrician's examination in the country. Consequently, this number should always be remembered.

Box Connections

Most boxes encountered will have removable knockouts for conduit and connector terminations. However, sometimes, cable or conduit must enter boxes that do not have pre-cut knockouts. In these cases, a knockout punch (Figure 6-11) may be used to make a hole for the conduit or connector.

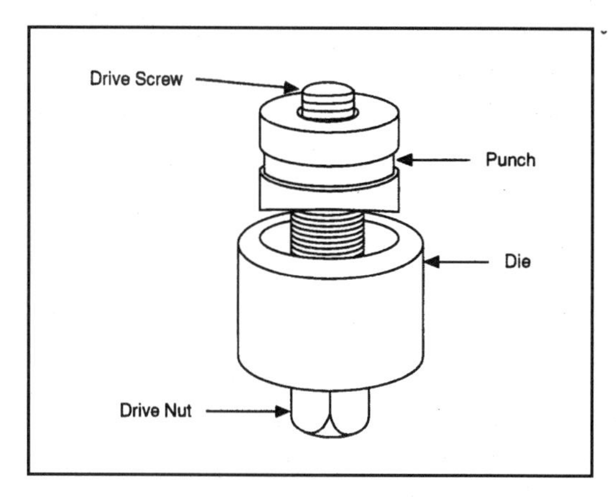

Figure 6-10: Required distances between conduit entries.

To use a knockout punch of the type shown in Figure 6-11, a hole must first be drilled through the box or cabinet. This hole must be large enough to accept the drive screw in the punch, then the punch and die are reassembled with the metal between the cutting edges. The drive nut is tightened, causing the punch to cut through the metal.

Bushings and Locknuts

Conduit is terminated at boxes by bushings and locknuts. Locknuts (Figure 6-12) secure the conduit or connector to the box while bushings protect the conductors from sharp edges. Bushings are usually

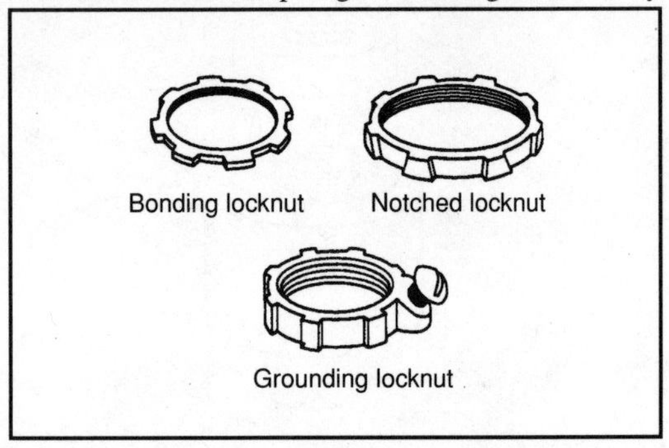

Figure 6-11: Basic parts of a knockout punch.

made of either fiber, plastic, or metal, and are available in several types as shown in Figure 6-13. Note that the grounding bushing has a lug with a set-screw for terminating the bonding wire.

Locknuts are frequently used on both the inside and outside walls of the enclosure to which the conduit terminates. However, if the bushing is metal and fits tightly against the inside wall of the enclosure, only the external locknut is needed on systems not exceeding 250 V.

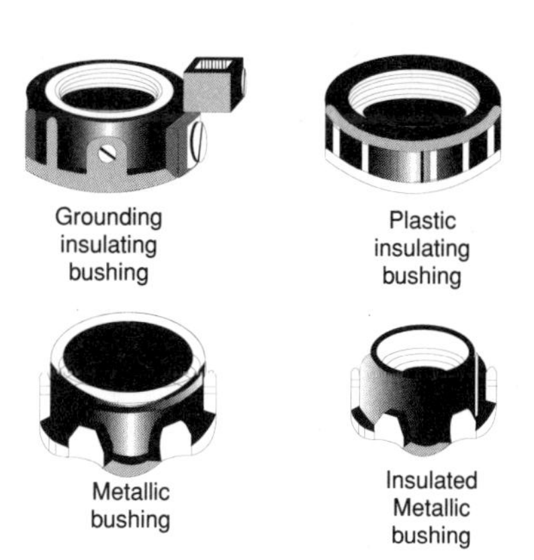

Figure 6-12: Various types of locknuts.

A conduit-to-box connection is shown in Figure 6-14. To make the connections as shown, use the following steps:

Step 1. Thread the external locknut onto the conduit end. Run the locknut to the bottom of the thread.

Step 2. Insert the conduit (with the locknut in place) into the box opening.

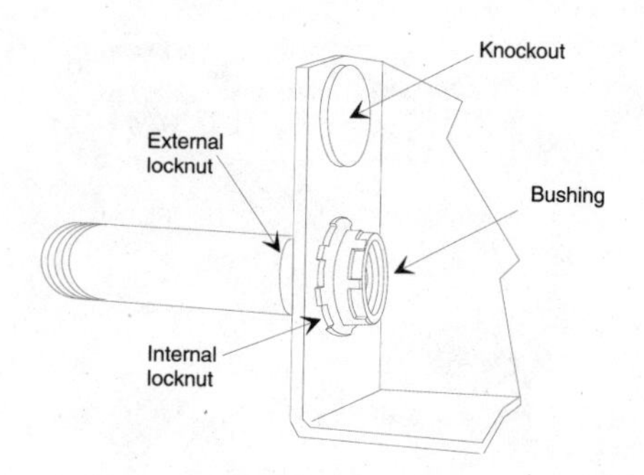

Figure 6-13: Various types of busings used on electrical installations.

Step 3. If an inside locknut or grounding locknut is required, screw it onto the conduit inside the box opening.

Step 4. Screw the bushing onto the threads projecting into the box opening. Make sure the bushing is adequately tightened.

Step 5. Tighten the external locknut to secure the conduit to the box.

It is important that the bushing and locknuts fit tightly against the box or cabinet. For this reason, the conduit must enter straight into the box. This may require an offset in the conduit to ensure a good fit.

Chapter 7

WIRING METHODS

Several types of wiring methods are used for electrical installations. The methods used on a given project are determined by several factors:

- The installation requirements set forth in the *National Electrical Code (NEC)*
- Local codes and ordinances
- Type of building construction
- Location of the wiring in the building
- Importance of the wiring system's appearance
- Costs and budget

In general, two types of basic wiring methods are used in the majority of electrical systems:

- Open wiring
- Concealed wiring

In open-wiring systems, the outlets and cable or raceway systems are installed on the surfaces of the walls, ceilings, columns, and the like where they are in view and readily accessible. Such wiring is often used in areas where appearance is not important and where it may be desirable to make changes in the electrical system at a later date. You will frequently find open-wiring systems in mechanical rooms and in interior parking areas of commercial buildings; the majority of wiring systems used in industrial applications will be exposed.

Concealed wiring systems have all cable and raceway runs concealed inside of walls, partitions, ceilings, columns, and behind baseboards or molding where they are out of view and not readily accessible. This type of wiring system is generally used in all new construction with finished interior walls, ceilings, floors and is the preferred type where good appearance is important.

Cable Systems

Several types of cable systems are used to construct electrical systems in various types of occupancies, and include the following:

Type NM Cable: This cable (Figure 7-1) is manufactured in two- or three-wire assemblies, and with varying sizes of conductors. In both two- and three-wire cables, conductors are color-coded: one conductor is black while the other is white in two-wire cable; in three-wire cable, the additional conductor is red. Both types will also have a grounding conductor which is usually bare, but is sometimes covered with a green plastic insulation — depending upon the

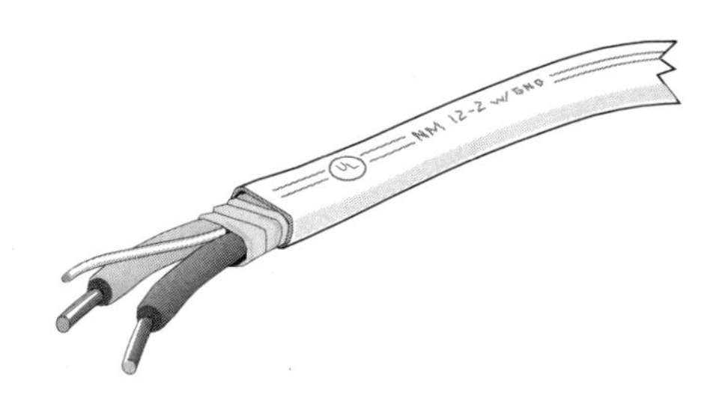

Figure 7-1: Type NM cable is often called "Romex" on the job.

manufacturer. The jacket or covering consists of rubber, plastic, or fiber. Most will also have markings on this jacket giving the manufacturer's name or trademark, the wire size, and the number of conductors. For example, "NM 12-2 W/GRD" indicates that the jacket contains two No. 12 AWG conductors along with a grounding wire; "NM 12-3 W/GRD" indicates three conductors plus a grounding wire. This type of cable may be concealed in the framework of buildings, or in some instances, may be run exposed on the building surfaces. It may not be used in any building exceeding three floors above grade; as a service-entrance cable; in commercial garages having hazardous locations; in theaters and similar locations; places of assembly; in motion picture studios; in storage battery rooms; in hoistways; embedded in poured concrete, or aggregate; or in any hazardous location except as otherwise permitted by the *NEC*.

Nonmetallic-sheathed cable is frequently referred to as *Romex* on the job.

Type AC (Armored) Cable: Type AC cable — commonly called "BX" — is manufactured in two-, three-, and four-wire assemblies, with varying sizes of conductors, and is used in locations similar to those where Type NM cable is allowed. The metallic spiral covering on BX cable offers a greater degree of mechanical protection than with NM cable, and the metal jacket also provides a continuous grounding bond without the need for additional grounding conductors. *See* Figure 7-2.

BX cable may be used for under-plaster extensions, as provided in the *NEC*, and embedded in plaster finish, brick, or other masonry, except in damp or wet locations. It may also be run or "fished" in the air voids of masonry block or tile walls, except where such walls are exposed or subject to excessive moisture or dampness or are below grade. This type of cable is a favorite for connecting 2 × 4 troffer-type lighting fixtures in commercial installations.

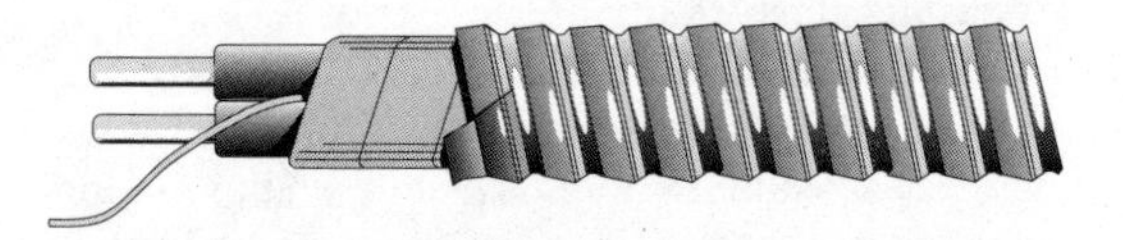

Figure 7-2: Type AC or "BX" cable.

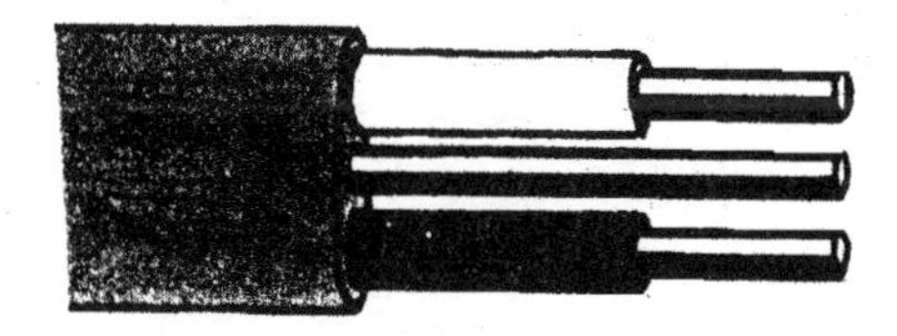

Figure 7-3: Underground feeder cable.

Underground Feeder Cable: Type UF cable (Figure 7-3) may be used underground, including direct burial in the earth, as a feeder or branch-circuit cable when provided with overcurrent protection at the rated ampacity as required by the *NEC*. When Type UF cable is used above grade where it will come in direct contact with the rays of the sun, its outer covering must be sun resistant. Furthermore, where Type UF cable emerges from the ground, some means of mechanical protection must be provided. This protection may be in the form of conduit or guard strips. Type UF cable resembles Type NM cable in appearance. The jacket, however, is constructed of weather resistant material to provide the required protection for direct-burial wiring installations.

Service-Entrance Cable: Type SE cable (Figure 7-4), when used for electrical services, must be installed as specified in *NEC* Article 230. This cable is available with the grounded conductor bare for out-

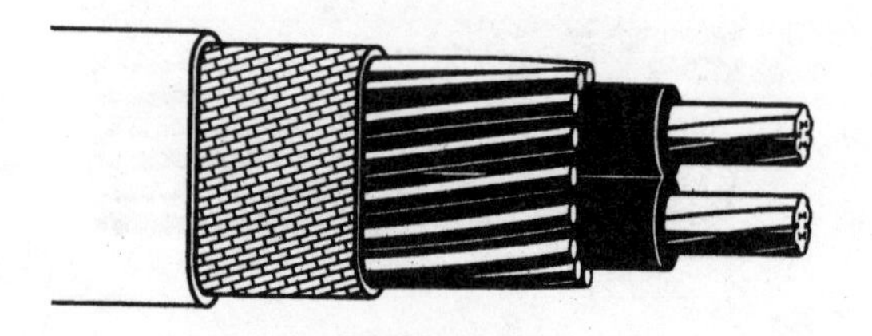

Figure 7-4: Type SE cable.

side service conductors, and also with an insulated grounded conductor (Type SER) for interior wiring systems.

Type SE and SER cable are permitted for use on branch circuits or feeders provided all current-carrying conductors are insulated; this includes the grounded or neutral conductor. When Type SE cable is used for interior wiring, all *NEC* regulations governing the installation of Type NM cable also apply to Type SE cable. There were some exceptions, but the 1996 *NEC* no longer permits Type SE cable with an uninsulated grounded conductor to be used on the following appliances:

- Electric range
- Wall-mounted oven
- Counter-mounted cooking unit
- Clothes dryer

Underground Service-Entrance Cable: Type USE cable is similar in appearance to Type SE cable except that it is approved for underground use and must be manufactured with a moisture-resistant covering. If a flame-retardant covering is not provided, it is not approved for indoor use.

Flat Conductor Cable: Type FCC cable consists of three or more flat copper conductors placed edge-to-edge and separated and enclosed within an insulating assembly. FCC systems consist of cable and associated shielding, connectors, terminators, adapters, boxes and receptacles. These systems are designed for installation under carpet squares on hard, sound, smooth, continuous floor surfaces made of concrete, ceramic, composition floor, wood, and similar materials. If used on heated floors with temperatures in excess of 86°F, the cable must be identified as suitable for use at these temperatures.

FCC systems must not be used outdoors or in corrosive locations; where subject to corrosive vapors; in any hazardous location; or in residential, school, or hospital buildings.

Flat-Cable Assemblies: This is Type FC cable assembly and should not be confused with Type FCC cable; there is a big difference. A Type FC wiring system is an assembly of parallel, special-stranded copper conductors formed integrally with an insulating material web specifically designed for field installation in surface metal raceway. The assembly is made up of three- or four-conductor cable, cable supports, splicers, circuit taps, fixture hangers, insulating end caps and other fittings. Guidelines for the use of this system are given in *NEC* Article 363. In general, the assembly is installed in an approved U-channel surface-metal raceway with one side open. Tap devices can be inserted anywhere along the channel. Connections from the tap devices to the flat-cable assembly are made by pin-type contacts when the tap devices are secured in place. The pin-type contacts penetrate the insulation of the cable assembly and contact the multistranded conductors in a matched phase sequence. These taps can then be connected to either lighting fixtures or power outlets.

Flat-cable assemblies must be installed for exposed work only and must not be installed in locations where they will be subjected to severe physical damage.

Mineral-Insulated Metal-Sheathed Cable: Type MI cable is a factory assembly of one or more conductors insulated with a highly compressed refractory mineral insulation and enclosed in a liquidtight and gas-tight continuous copper sheath. It may be used for electric services, feeders, and branch circuits in dry, wet, or continuously moist locations. Furthermore, it may be used indoors or outdoors, embedded in plaster, concrete, fill, or other masonry, whether above or below grade. This type of cable may also be used in hazardous locations, where exposed to oil or gasoline, where exposed to corrosive conditions not deteriorating to the cable's sheath, and in underground runs where suitably protected against physical damage and corrosive conditions. In other words, MI cable may be used in practically any electrical installation.

Power and Control Tray Cable: Type TC power and control tray cable is a factory assembly of two or more insulated conductors, with or without associated bare or covered grounding conductors, under a nonmetallic sheath, approved for installation in cable trays, in raceways, or where supported by a messenger wire. The use of this cable is limited to commercial and industrial applications where the conditions of maintenance and supervision assure that only qualified persons will service the installation.

Metal-Clad Cable: Type MC cable is a factory assembly of one or more conductors, each individually insulated and enclosed in a metallic sheath of interlocking tape or a smooth or corrugated tube. This type of cable may be used for services, feeders, and branch circuits; power, lighting, control, and signal circuits; indoors or outdoors; where exposed or concealed; direct buried; in cable tray; in any approved raceway; as open runs of cable; as aerial cable on a messenger; in hazardous locations as permitted in *NEC* Articles 501, 502, and 503; in dry locations; and in wet locations under certain conditions as specified in the *NEC*.

Raceway Systems

A raceway wiring system consists of an electrical wiring system in which one or more individual conductors are pulled into a conduit or similar housing, usually after the raceway system has been completely installed. The basic raceways are rigid steel conduit, electrical metallic tubing (EMT), and PVC (polyvinyl chloride) plastic. Other raceways include surface metal moldings and flexible metallic conduit.

These raceways are available in standardized sizes and serve primarily to provide mechanical protection for the wires run inside and, in the case of metallic raceways, to provide a continuously grounded system. Metallic raceways, properly installed, provide the greatest degree of mechanical and grounding protection and provide maximum protection against fire hazards for the electrical system. However, they are more expensive to install.

Most electricians prefer to use a hacksaw with a blade having 18 teeth per inch for cutting rigid conduit and 32 teeth per inch for cutting the smaller sizes of conduit. For cutting larger sizes of conduit ($1\frac{1}{2}$ inches and above), a special conduit cutter should be used to save time. While quicker to use, the conduit cutter almost always leaves a hump inside the conduit and the burr is somewhat larger than made by a standard hacksaw. If a power band saw is available on the job, it is preferred for cutting the larger sizes of conduit. Abrasive cutters are also popular for the larger sizes of conduit.

Conduit cuts should be made square and the inside edge of the cut must be reamed to remove any burr or sharp edge that might damage wire insulation when the conductors are pulled inside the conduit. After reaming, most experienced electricians feel the inside of the cut with their finger to be sure that no burrs or sharp edges are present.

Lengths of conduit to be cut should be accurately measured for the size needed and an additional $\frac{3}{8}$ inch should be allowed on the smaller sizes of conduit for terminations; the larger sizes of conduit will require approximately $\frac{1}{2}$ inch for locknuts, bushings, and the like at terminations.

A good lubricant (cutting oil) is then used liberally during the thread-cutting process. If sufficient lubricant is used, cuts may be made cleaner and sharper, and the cutting dies will last much longer.

Full threads must be cut to allow the conduit ends to come close together in the coupling or to firmly seat in the shoulders of threaded hubs of conduit bodies. To obtain a full thread, run the die up on the conduit until the conduit barely comes through the die. This will give a good thread length adequate for all purposes. Anything longer will not fit into the coupling and will later corrode because threading removes the zinc or other protective coating from the conduit.

Clean, sharply cut threads also make a better continuous ground and save much trouble once the system is in operation.

Electrical Metallic Tubing

Electrical metallic tubing (EMT) may be used for both exposed and concealed work except where it will be subjected to severe damage during use, in cinder concrete, or in fill where subjected to permanent moisture unless some means to protect it is provided; the tubing may be installed a minimum of 18 inches under the fill.

Threadless couplings and connectors are used for EMT installation and these should be installed so that the tubing will be made up tight. Both set-screw and compression types are commonly in use. Where buried in masonry or installed in wet locations, couplings and connectors, as well as supports, bolts, straps, and screws, should be of a type approved for the conditions.

Bends in the tubing should be made with a tubing bender so that no injury will occur and so the internal diameter of the tubing will not be effectively reduced. The bends between outlets or termination points should contain no more than the equivalent of four quarter-bends (360° total), including those bends located immediately at the outlet or fitting (offsets).

All cuts in EMT are made with either a hacksaw, power hacksaw, tubing cutter, or other approved device. Once out, the tubing ends should be reamed with a screwdriver handle or pipe reamer to remove all burrs and sharp edges that might damage conductor insulation.

Flexible Metal Conduit

Flexible metal conduit generally is manufactured in two types, a standard metal-clad type and a liquid-tight type. The former type cannot be used in wet locations unless the conductors pulled in are of a type specially approved for such conditions. Neither type may be used where they will be subjected to physical damage or where any combination of ambient and/or conductor temperature will produce an operating temperature in excess of that for which the material is approved. Other uses are fully described in Articles 350 and 351 of the *NEC*.

When this type of conduit is installed, it should be secured by an approved means at intervals not exceeding 4½ feet and within 12 inches of every outlet box, fitting, or other termination points. In some cases, however, exceptions exist. For example, when flexible metal conduit must be finished in walls, ceilings, and the like, securing the conduit at these intervals would not be practical. Also, where more flexibility is required, lengths of not more than 3 feet may be utilized at termination points.

Flexible metal conduit may be used as a grounding means where both the conduit and the fittings are approved for the purpose. In lengths of more than 6 feet, it is best to install an extra grounding conductor within the conduit for added insurance.

Liquidtight flexible metal conduit is used in damp or wet locations and is covered in *NEC* Article 351. Please see the *NEC* book for further details.

Surface Metal Molding

When it is impractical to install the wiring in concealed areas, surface metal molding is a good compromise. Even though it is visible, proper painting to match the color of the ceiling and walls makes it very inconspicuous. Surface metal molding is made from sheet metal strips drawn into shape and comes in various shapes and sizes with factory fittings to meet nearly every application found in finished areas of commercial buildings. A complete list of fittings can be obtained at your local electrical equipment supplier.

The running of straight lines of surface molding is simple. A length of molding with the coupling is slipped in the end, out enough so that the screw hole is exposed, and then the coupling is screwed to the surface to which the molding is to be attached. Then another length of molding is slipped on the coupling.

Factory fittings are used for corners and turns or the molding may be bent (to a certain extent) with a special bender. Matching outlet boxes for surface mounting are also available, and bushings are necessary at such boxes to prevent the sharp edges of the molding from injuring the insulation on the wire.

Clips are used to fasten the molding in place. The clip is secured by a screw and then the molding is slipped into the clip, wherever extra support of the molding is needed, and fastened by screws. When parallel runs of molding are installed, they may be secured in place by means of a multiple strap. The joints in runs of molding are covered by slipping a connection cover over the joints. Such runs of molding should be grounded the same as any other metal raceway, and this is done by use of grounding clips. The current-carrying wires are normally pulled in after the molding is in place.

The installation of surface metal molding requires no special tools unless bending the molding is necessary. The molding is fastened in place with screws, toggle bolts, and the like, depending on the materials to which it is fastened. All molding should be run straight and parallel with the room or building lines, that is, baseboards, trims, and other room moldings. The decor of the room should be considered first and the molding made as inconspicuous as possible.

It is often desirable to install surface molding not used for wires in order to complete a pattern set by other surface molding containing current-carrying wires, or to continue a run to make it appear to be part of the room's decoration.

Wireways

Wireways are sheet-metal troughs with hinged or removable covers for housing and protecting wires and cables and in which conductors are held in place after the wireway has been installed as a complete system. They may be used only for exposed work and shouldn't be installed where they will be subject to severe physical damage or corrosive vapor nor in any hazardous location except Class II, Division 2 of the *NEC*.

The wireway structure must be designed to safely handle the sizes of conductors used in the system. Furthermore, the system should not contain more than 30 current-carrying conductors at any cross section. The sum of the cross-sectional areas of all contained conductors at any cross section of a wireway shall not exceed 20 percent of the interior cross-sectioned area of the wireway.

Splices and taps, made and insulated by approved methods, may be located within the wireway provided they are accessible. The conductors, including splices and taps, shall not fill the wireway to more than 75 percent of its area at that point.

Wireways must be securely supported at intervals not exceeding 5 feet, unless specially approved for supports at greater intervals, but in no case shall the distance between supports exceed 10 feet.

Busways

There are several types of busways or duct systems for electrical transmission and feeder purposes as shown in Figures 7-5 through 7-8. Lighting duct, trolley duct, and distribution bus duct are just a few. All are designed for a specific purpose, and electricians should become familiar with all types before an installation is laid out.

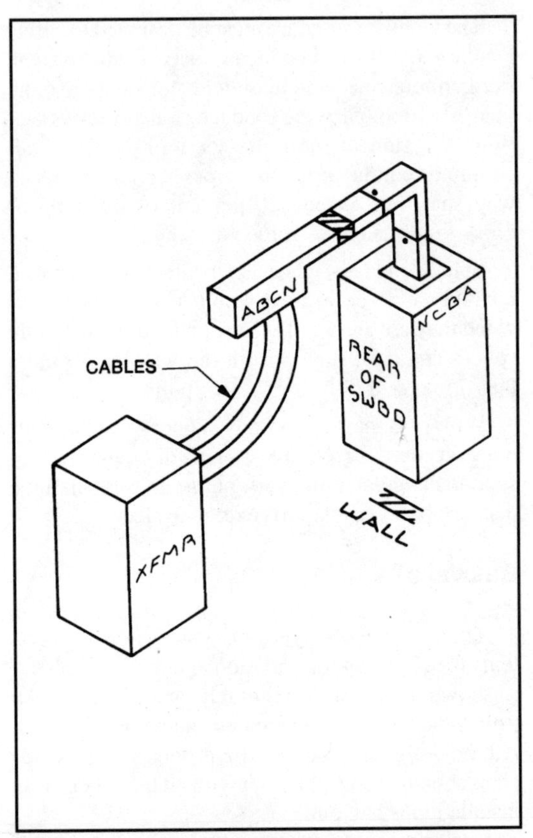

Figure 7-5: Typical busway run from a utility transformer to a switchboard.

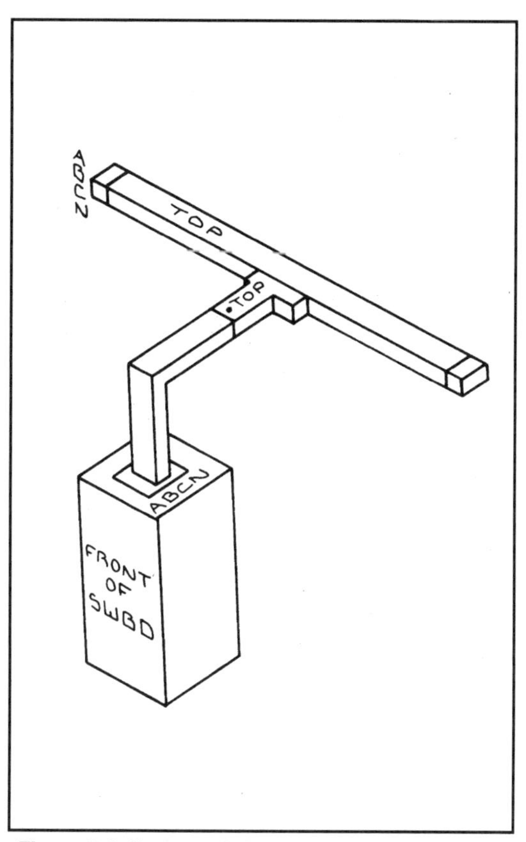

Figure 7-6: Horizontal plug-in busway using a plug-in tee.

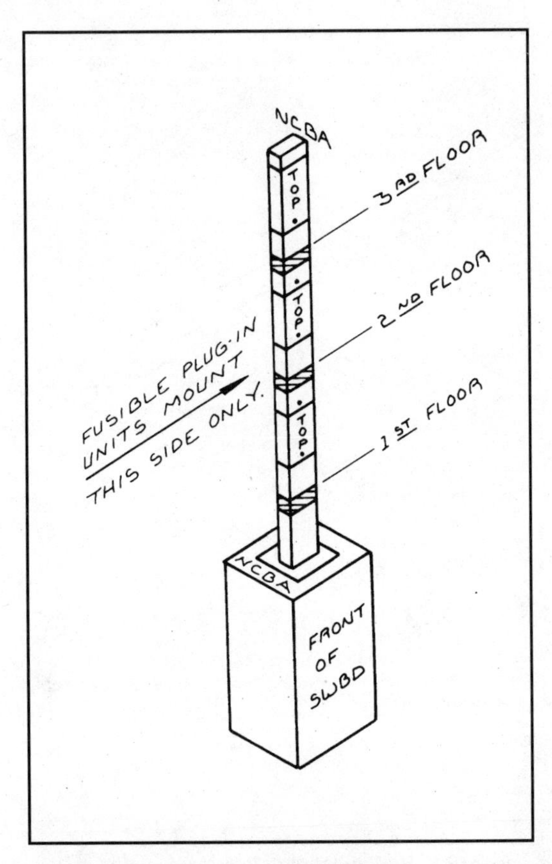

Figure 7-7: Plug-in type vertical riser.

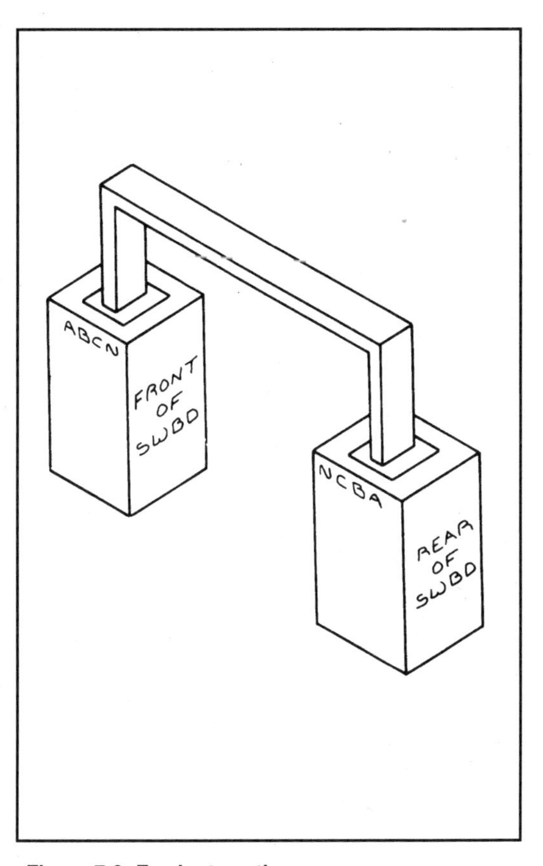

Figure 7-8: Feeder type tie run.

Lighting duct, for example, permits the installation of an unlimited amount of footage from a single working platform. As each section and the lighting fixtures are secured in place, the complete assembly is then simply transported to the area of installation and installed in one piece.

Trolley duct is widely used for industrial applications, and where the installation requires a continuous polarization to prevent accidental reversal, a polarizing bar is used. This system provides polarization for all trolley, permitting standard and detachable trolleys to be used on the same run.

Plug-in bus duct is also widely used for industrial applications, and the system consists of interconnected prefabricated sections of bus duct so formed that the complete assembly will be rigid in construction and neat and symmetrical in appearance.

Cable Trays

Cable trays are used to support electrical conductors used mainly in industrial applications, but are sometimes used for communication and data processing conductors in large commercial establishments. The trays themselves are usually made up into a system of assembled, interconnected sections and associated fittings, all of which are made of metal or other noncombustible material. The finished system forms into a rigid structural run to contain and support single, multiconductor, or other wiring cables. Several styles of cable trays are available, including ladder, trough, channel, solid-bottom trays, and similar structures.

Underfloor Raceway

Underfloor raceway consists of ducts laid below the surface of the floor and interconnected by means of fittings and outlet or junction boxes. Both metallic and nonmetallic ducts are used. Obviously, this system must be installed prior to the floor being finished.

Identifying Conductors

The *NEC* specifies certain methods of identifying conductors used in wiring systems of all types. For example, the high leg of a 120/240-volt grounded three-phase delta system must be marked with an orange color for identification; a grounded conductor must be identified either by the color of its insulation, by markings at the terminals, or by other suitable means. Unless allowed by *NEC* exceptions, a grounded conductor must have a white or natural gray finish. When this is not practical for conductors larger than No. 6 AWG, marking the terminals with white color is an acceptable method of identifying the conductors.

Color Coding

Conductors contained in cables are color-coded so that identification may be easily made at each access point. The following lists the color-coding for cables up through five-wire cable:

- Two-conductor cable: one white wire, one black wire, and a grounding conductor (usually bare)
- Three-conductor cable: one white, one black, one red, and a grounding conductor
- Four-conductor cable: fourth wire blue
- Five-conductor cable: fifth wire yellow
- The grounding conductor may be either green or green with yellow stripes

Although some control-wiring and communication cables contain 60, 80, or more pairs of conductors — using a combination of colors — the ones listed are the most common and will be encountered the most on electrical installations.

When conductors are installed in raceway systems, any color insulation is permitted for the ungrounded phase conductors except the following:

White or gray	Reserved for use as the grounded circuit conductor
Green	Reserved for use as a grounding conductor only

Changing Colors

Should it become necessary to change the actual color of a conductor to meet *NEC* requirements or to facilitate maintenance on circuits and equipment, the conductors may be reidentified with colored tape or paint.

For example, assume that a two-wire cable containing a black and white conductor is used to feed a 240-volt, two-wire single-phase motor. Since the white colored conductor is supposed to be reserved for the grounded conductor, and none is required in this circuit, the white conductor may be marked with a piece of red tape at each end of the circuit so that everyone will know that this wire is not a grounded conductor.

Conductors in Conduit

In most cases, the installation of conductors in raceway systems is merely routine for the experienced electrician. However, there are certain practices that can reduce labor, materials, and help prevent damage to the conductors. The use of modern equipment, such as vacuum fish-tape systems, is one way to reduce labor during this phase of the wiring installation. The proper size and length of the fish tape, as well as the type, should be one of the first considerations. For example, if most of the runs between branch-circuit outlets are only 20 feet or less, a short fish tape of, say, 25 feet will easily handle the job and will not have the weight and bulk of a larger one. When longer runs are encountered the required length of the fish tape should be enclosed in one of the metal or plastic fish-tape reels. This way the fist tape can be rewound on the reel as the pull is being made to avoid having an excessive length of tape lying around on the floor or deck.

When several bends are present in the raceway system, the insertion of the fish tape may be made easier by using flexible fish-tape leaders on the end of the fish tape.

The combination blower and vacuum fish-tape systems are ideal for use on long runs and can save much time. Basically, the system consists of a tank and air pump with accessories. An electrician can vacuum or blow a line or tape in any size conduit from ½ through 4 inches, or even up to 6-inch conduit with optional accessories.

After the fish tape is inserted in the raceway system, the conductors must be firmly attached by some approved means. On short runs, where only a few conductors are involved, all that is necessary is to strip the insulation from the ends of the wires, bend these ends around the hook in the fish tape, and securely tape them in place. Where several wires are to be pulled together, the wires should be staggered and the fish tape securely taped at the point of attachment so that the overall diameter is not increased any more than is absolutely necessary. Staggering is done by attaching one wire to the fish tape and then attaching the second wire a short distance behind this to the bare conductor of the first wire. The third wire, in turn, is attached to the second wire and so forth as shown in Figure 7-9.

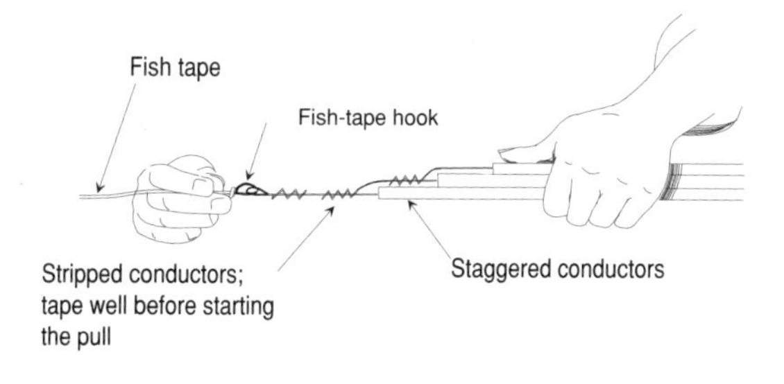

Figure 7-9: Staggering conductors on a fish tape for short pulls in conduit.

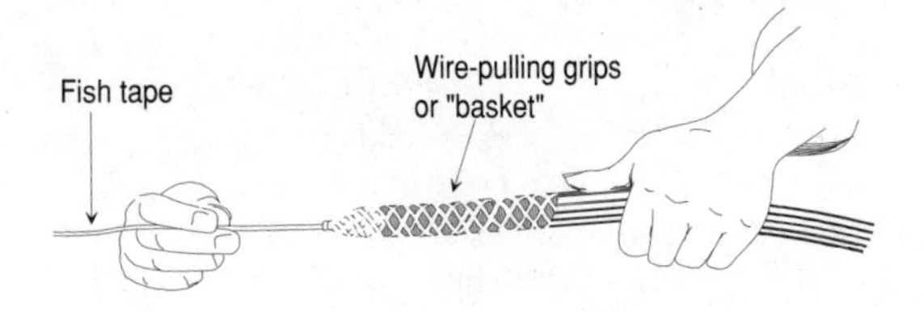

Figure 7-10: Basket grips are frequently used for both hand and power pulls.

Basket grips (Figure 7-10) are available in many sizes for almost any size and combination of conductors. They are designed to hold the conductors firmly to the fish tape and can save the electrical workers much time and trouble that would be required when taping wires.

In all but very short runs, the wires should be lubricated with a good quality wire lubricant prior to attempting the pull, and also during the pull. Some of this lubricant should also be applied to the inside of the conduit itself.

Wire dispensers are great aids in keeping the conductors straight and facilitating pulling of conductors. Many different types of wire dispensers are now marketed that handle virtually any size spool of wire

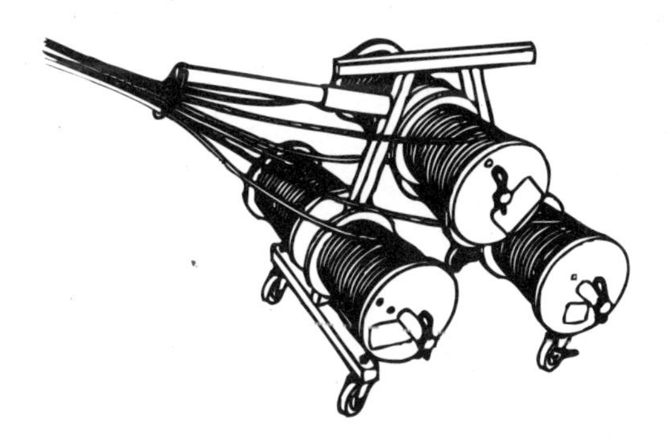

Figure 7-11: Conductor dispenser used for wire pulling.

or cable. Some of the smaller dispensers can handle up to ten spools of wire from No. 22 to No. 10 AWG; the larger ones can handle a lot more. These dispensers are sometimes called *wire caddies*. See Figure 7-11.

Planning the Installation

The importance of planning any wire-pulling installation cannot be stressed too much. Proper planning will make the work go easier and much labor will always be saved.

Large sizes of conductors are usually shipped on reels, involving considerable weight and bulk. Consequently, setting up these reels for the pull, measuring cable run lengths, and similar preliminary steps will often involve a relatively large amount of the total cable installation time. Therefore, consideration must be given to reel set-up space, proper equipment, and moving the cable reels into place.

Whenever possible, the conductors should be pulled directly from the shipping reels without prehandling them. This can usually be done through proper coordination of the ordering of the conductors with the job requirements. While doing so requires extremely close checking of the drawings and on-the-job measurements, allowing for adequate lengths of conductors in pull boxes, elbows and troughs, and for connections and splices, the extra effort is well worth the time to all involved.

When the lengths of cable have been established, the length of cable per reel can be ordered so that the total length per reel will be equal to the total of a given number of raceway lengths, and the reel so identified.

In most cases, the individual cables of the proper length for a given number of runs are reeled separately onto two or more reels at the factory, depending upon the number of conductors in the runs.

When individual conductors are shipped on separate reels, it is necessary to set up for the same number of reels as the number of conductors to be pulled into a given run.

As an extra precaution against error in calculating the lengths of conductors involved, it is good practice to actually measure all runs with a fish tape before starting the cable pull, checking the totals against the totals indicated on the reels. Under normal cable delivery schedules, when the feeder raceways have been installed at a relatively early stage of the overall building construction, it may not delay the final completion of the electrical installation to delay ordering the cables until the raceways can actually be measured.

In pulling conductors directly from the reels, care must be taken that each given run be cut off from the reel so that there is a minimum amount of waste. In other words, preclude the possibility of the final run of cable taken from the reel being too short for that run.

Pulling Location

Each job will have to be judged separately as to the best location of pulling setups; the number of setups should be reduced to a minimum in line with the best direction of pull. For example, it is usually best to pull cables downward rather than up, to avoid having to pull the total weight of the cable at the final stages of the pull and to avoid the possibility of injury to workers should the conductors break loose from the pulling cable on long vertical pulls. On the other hand, it may not be practical to hoist the cable reels and setup equipment to the upper locations in the building. Also, a separate setup might have to be made at the top of each rise, whereas a single setup might be made at a ground floor pull box location from which several feeders were served with the same size and number of conductors.

The location of the pulling equipment determines the number of workers required for the job. A piece of equipment that can be moved in and set up on the first floor by four workers in an hour's time may require six workers working two hours when set up in basements or parking levels of buildings.

Every employer expects a day's work for a day's pay, but no employer should expect workers to do the work of two or three. Therefore, when planning a cable pull, make certain enough workers are on hand to adequately handle the installation.

It is a simple operation for a few workers to roll cable reels from a loading platform to a first floor setup, whereas moving them to upper floors involves much handling and usually requires a crane or other hoists. In addition, the reel jacks, and sometimes braking equipment have to be moved to the setup point when a downward pull is made. After the pulling operation is completed, reels, jacks and cable brakes eventually all have to be taken back to the first floor.

Cable Pull Operations

The following operations are performed to a lesser or greater degree in almost all cable pulls with larger sizes of conductors:

Step 1 Measuring or rechecking runs.

Step 2 Providing pulling equipment.

Step 3 Receiving and unloading pulling equipment.

Step 4 Moving pulling equipment to the pulling location.

Step 5 Setting and anchoring pulling equipment.

Step 6 Removing to another location or moving to a loading platform and loading on truck or forklift.

Step 7 Receiving and storing cable; may be moved directly to setup location if job conditions permit.

Step 8 Moving to setup point.

Step 9 Moving reel jacks and mandrel to setup point.

Step 10 Jacking up reels.

Step 11 Preparing cable ends for the pull.

Step 12 Installing fish tape.

Step 13 Installing pulling line or cable.

Step 14 Connecting pulling line.

Step 15 Lubricating with proper lubricants.

Step 16 Pulling cable.

Step 17 Disconnecting pulling line.

Step 18 Removing reels.

Step 19 Identifying conductor "legs."

Step 20 Racking cables in pull boxes and troughs.

Step 21 Splicing or connecting cables.

Step 22 Checking and testing.

In some instances the following additional operations are involved, depending upon the exact details of the project.

Step 23 Removing lagging from reels.

Step 24 Unreel cable and cut to length.

Step 25 Re-reel cable for pulling.

Step 26 Replacing lagging on reels.

Step 27 Operating such auxiliary equipment as cable brakes, guiding through cabinets or pull boxes, signaling between reel and pulling setups.

The knowledge gained from this information should enable the apprentice electrician to become one step closer to becoming a first-class electrician. But your education does not stop here. Rather, workers gain new knowledge about the electrical industry almost every day on the job. This applies to both apprentice and experienced workers alike. You must apply this knowledge to practical on-the-job applications to get the full benefit.

Size	Temperature Rating of Conductors		
	60°C (140°F)	75°C (167°F)	90°C (194°F)
AWG kcmil	TYPES TW†, UF†	TYPES FEPW†, RH†, RHW†, THHW†, THW†, THWN†, XHHW† USE†, ZW†	TYPES TA, TBS, SA SIS, FEP†, FEPB†, MI RHH†, RHW-2, THHN†, THHW†, THW-2, THWN-2, USE-2, XHH, XHHW† XHHW-2, ZW-2
	COPPER		
18	....	....	14
16	....	....	18
14	20†	20†	25†
12	25†	25†	30†
10	30	35†	40†
8	40	50	55
6	55	65	75
4	70	85	95
3	85	100	110
2	95	115	130
1	110	130	150
1/0	125	150	170
2/0	145	175	195
3/0	165	200	225
4/0	195	230	260
250	215	255	290
300	240	285	320
350	260	310	350
400	280	335	380
500	320	380	430
600	355	420	475
700	385	460	520
750	400	475	535
800	410	490	555
900	435	520	585
1000	455	545	615
1250	495	590	665
1500	520	625	705
1750	545	650	735
2000	560	665	750

Figure 7-12: Allowable ampacities of insulated copper conductors (0 - 2000V) with not more than 3 conductors per raceway.

Temperature Rating of Conductors			Size
60°C (140°F)	75°C (167°F)	90°C (194°F)	
TYPES TW†, UF†	TYPES RH†, RHW†, THHW†, THW†, THWN†, XHHW†, USE†	TYPES TA, TBS, SA, SIS, THHN†, THHW†, THW-2, THWN-2, RHH†, RHW-2 USE-2 XHH, XHHW XHHW-2, ZW-2	AWG kcmil
ALUMINUM OR COPPER-CLAD ALUMINUM			
....	....	....	....
....	....	....	....
....	....	....	....
20†	20†	25†	12
25	30†	35†	10
30	40	45	8
40	50	60	6
55	65	75	4
65	75	85	3
75	90	100	2
85	100	115	1
100	120	135	1/0
115	135	150	2/0
130	155	175	3/0
150	180	205	4/0
170	205	230	250
190	230	255	300
210	250	280	350
225	270	305	400
260	310	350	500
285	340	385	600
310	375	420	700
320	385	435	750
330	395	450	800
355	425	480	900
375	445	500	1000
405	485	545	1250
435	520	585	1500
455	545	615	1750
470	560	630	2000

Figure 7-13: Allowable ampacities of insulated aluminum or copper-clad conductors (0 - 2000V) with not more than 3 conductors per raceway.

Notes to Tables

Unless otherwise specifically permitted elsewhere in the *NEC*, the overcurrent protection for conductor types marked with an obelisk (†) in Figures 7-12 and 7-13 shall not exceed 15 A for No 14, 20 A for No. 12, and 30 A for No. 10 copper; or 15 A for No 12 and 25 A for No. 10 aluminum and copper-clad aluminum after any correction factors for ambient temperature and number of conductors have been applied.

Chapter 8

WIRING DEVICES

A *device* — by *NEC* definition — is a unit of an electrical system that is intended to carry, but not utilize, electric energy. This covers a wide assortment of system components that include, but are not limited to, the following:

- Switches
- Relays
- Contactors
- Receptacles
- Conductors

However, for our purpose, we will deal with switching devices and also those items used to connect utilization equipment to electrical circuits; namely, *receptacles*. Both are commonly known as *wiring devices*.

Receptacles

A receptacle is a contact device installed at the outlet for the connection of a single attachment plug. Several types and configurations are available for use with many different attachment plug caps — each designed for a specific application. For example, receptacles are available for two-wire120-V 15- and 20-A circuits; others are designed for use on two- and three-wire, 240-volt, 20-, 30-, 40-, and 50-A circuits. There are also many other types, some of which are discussed in this chapter.

Receptacles are rated according to their voltage and amperage capacity. This rating, in turn, determines the number and configuration of the contacts — both on the receptacle and the receptacle's mating plug. Figure 8-1 shows a few of the more common receptacle configurations.

Receptacle Characteristics

Receptacles have various symbols and information inscribed on them that help to determine their proper use and ratings. For example, Figure 8-2 shows a standard duplex receptacle and contains the following printed inscriptions:

- The testing laboratory label
- The CSA (Canadian Standards Association) label
- Type of conductor for which the terminals are designed
- Current and voltage ratings, listed by maximum amperage, maximum voltage, and current restrictions

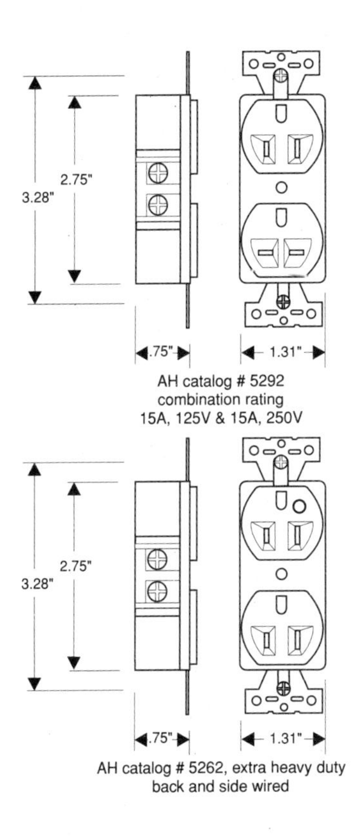

Figure 8-1: NEMA configurations for general-purpose nonlocking receptacles and plug caps.

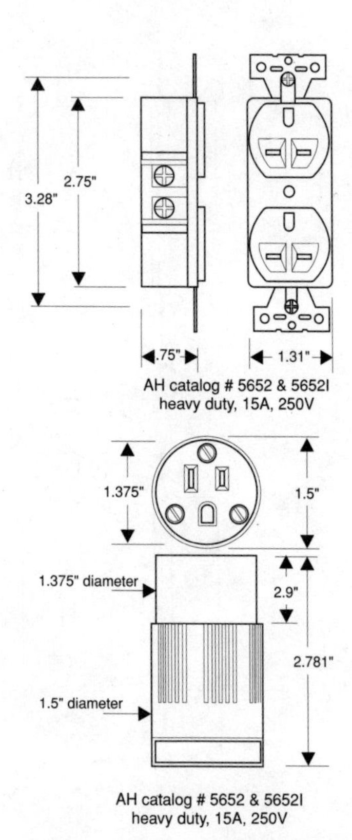

Figure 8-1: NEMA configurations for general-purpose nonlocking receptacles and plug caps. ***(Cont.)***

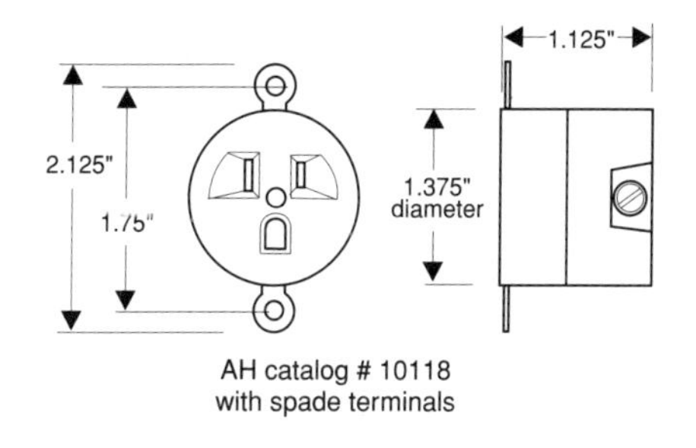

AH catalog # 10118
with spade terminals

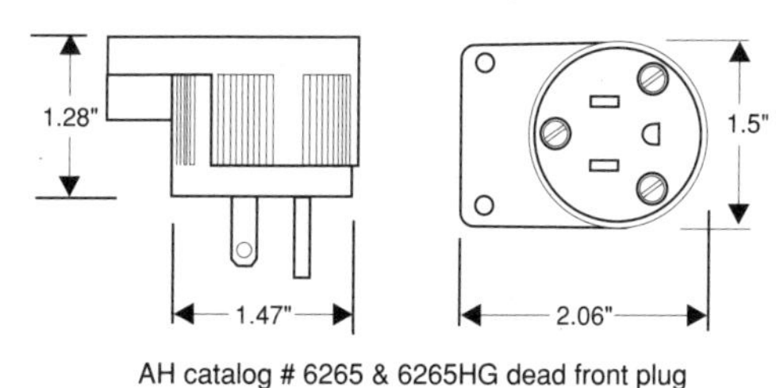

AH catalog # 6265 & 6265HG dead front plug
angle style, back-wired, 15A, 125V

Figure 8-1: NEMA configurations for general-purpose nonlocking receptacles and plug caps. ***(Cont.)***

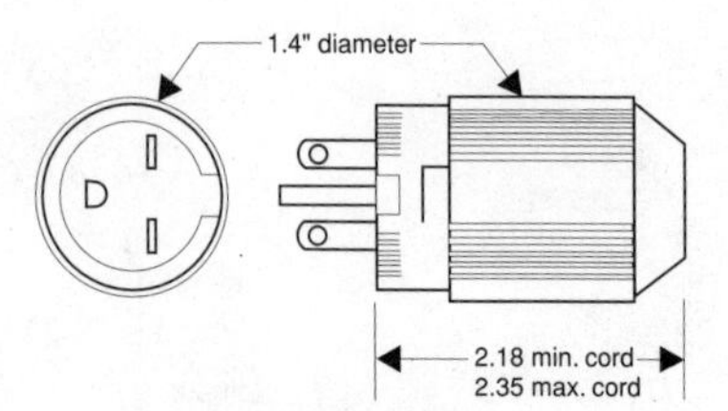

AH catalog # 5666V dead front plug,
2-pole,3-wire grounding15A, 125V

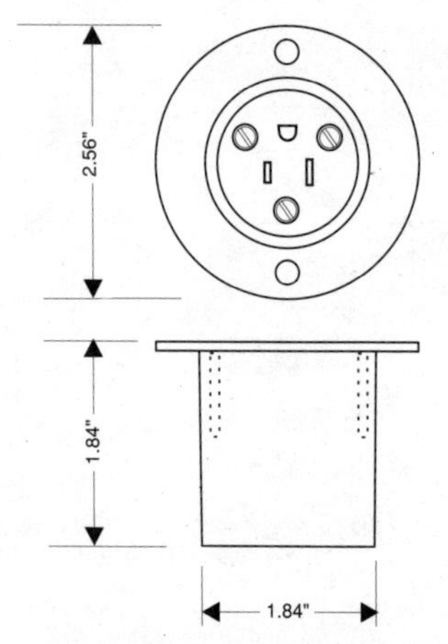

AH catalog # 5279C flanged outlet
2-pole,3-wire grounding 15A, 125V

Figure 8-1: NEMA configurations for general-purpose nonlocking receptacles and plug caps. *(Cont.)*

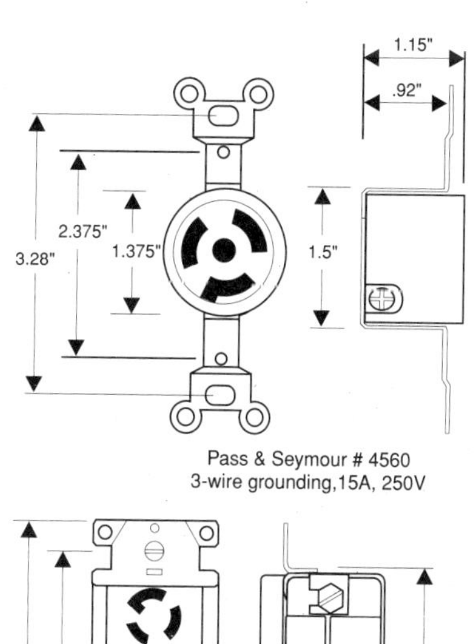

Pass & Seymour # 4560
3-wire grounding,15A, 250V

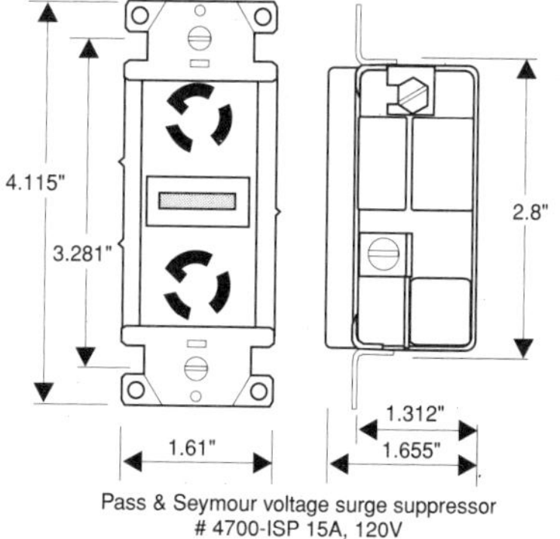

Pass & Seymour voltage surge suppressor
4700-ISP 15A, 120V

Figure 8-1: NEMA configurations for general-purpose receptacles and plug caps. ***(Cont.)***

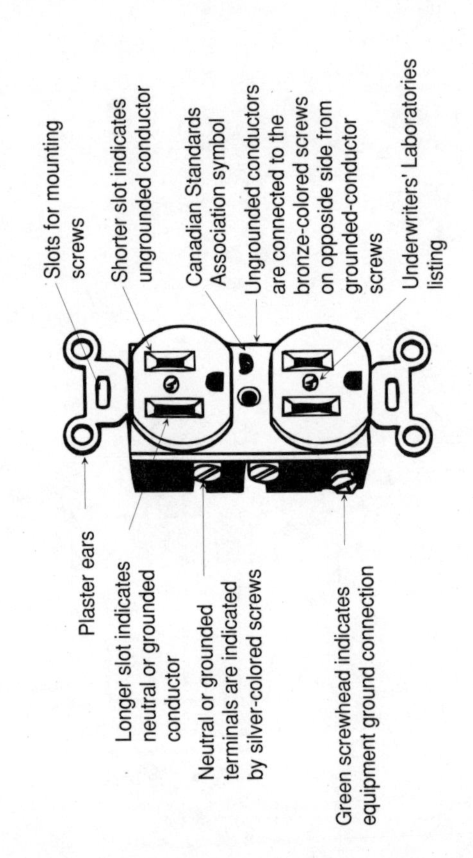

Figure 8-2: Characteristics of a typical duplex receptacle.

The testing laboratory label is an indication that the device has undergone extensive testing by a nationally recognized testing lab and has met with the minimum safety requirements. The label does not indicate any type of quality rating. The receptacle in Figure 8-2 is marked with the "UL" label which indicates that the device type was tested by Underwriters' Laboratories, Inc. of Northbrook, IL. ETL Testing Laboratories, Inc. of Cortland, NY, is another nationally recognized testing laboratory. They provide a labeling, listing and follow-up service for the safety testing of electrical products to nationally recognized safety standards or specifically designated requirements of jurisdictional authorities.

The CSA (Canadian Standards Association) label is an indication that the material or device has undergone a similar testing procedure by the Canadian Standards Association and is acceptable for use in Canada.

Current and voltage ratings are listed by maximum amperage, maximum voltage and current restriction. On the device shown in Figure 8-2, the maximum current rating is 15 A at 125 V — the latter of which is the maximum voltage allowed on a device so marked.

Conductor markings are also usually found on duplex receptacles. Receptacles with quick-connect wire clips will be marked "Use #12 or #14 solid wire only." If the inscription "CO/ALR" is marked on the receptacle, either copper, aluminum, or copper-clad aluminum wire may be used. The letters "ALR" stand for "aluminum revised." Receptacles marked with the inscription "CU/AL" should be used for copper only, although they were originally intended for use with aluminum also. However, such devices frequently failed when connected to 15- or 20-A circuits. Consequently, devices marked with "CU/AL" are no longer acceptable for use with aluminum conductors.

The remaining markings on duplex receptacles may include the manufacturer's name or logo, "Wire Release" inscribed under the wire-release slots, and the letters "GR" beneath or beside of the green grounding screw.

The screw terminals on receptacles are color-coded. For example, the terminal with the green screw head is the equipment ground connection and is connected to the U-shaped slots on the receptacle. The silver-colored terminal screws are for connecting the grounded or neutral conductors and are associated with the longer of the two vertical slots on the receptacle. The brass-colored terminal screws are for connecting the ungrounded or "hot" conductors and are associated with the shorter vertical slots on the receptacle.

Note:

> *The long vertical slot on receptacles accepts the grounded or neutral conductor while the shorter vertical slot accepts the ungrounded or hot conductor.*

Switches

The purpose of a switch is to make and break an electrical circuit, safely and conveniently. In doing so, a switch may be used to manually control lighting, motors, fans, and other various items connected to an electrical circuit. Switches may also be activated by light, heat, chemicals, motion, and electrical energy for automatic operation. *NEC* Article 380 covers the installation and use of switches.

Although there is some disagreement concerning the actual definitions of the various switches that might fall under the category of *wiring devices*, the most generally accepted ones are as follows:

Bypass isolation switch: This is a manually operated device used in conjunction with a transfer switch to provide a means of directly connecting load conductors to a power source, and of disconnecting the transfer switch.

General-use switch: A switch intended for use in general distribution and branch circuits. It is rated in amperes, and it is capable of interrupting its rated current at its rated voltage.

General-use snap switch: A form of general-use switch so constructed that it can be installed in flush device boxes or on outlet-box covers, or otherwise used in conjunction with wiring systems recognized by the *NEC*.

Isolating switch: A switch intended for isolating an electric circuit from the source of power. It has no interrupting rating, and it is intended to be operated only after the circuit has been opened by some other means.

Motor-circuit switch: A switch, rated in horsepower, capable of interrupting the maximum operating overload current of a motor of the same horsepower rating as the switch at its rated voltage.

Transfer switch: A transfer switch is a device for transferring one or more load conductor connections from one power source to another. This type of switch may be either automatic or nonautomatic.

Common Switch Terms

A brief review of these terms is warranted here. In general, the major terms used to identify the characteristics of switches are:

- Pole or poles
- Throw

The term *pole* refers to the number of conductors that the switch will control in the circuit. For example, a single-pole switch breaks the connection on only one conductor in the circuit. A double-pole switch breaks the connection to two conductors, and so forth.

The term *throw* refers to the number of internal operations that a switch can perform. For example, a single-pole, single-throw switch will "make" one conductor when thrown in one direction — the "ON" direction — and "break" the circuit when thrown in the opposite direction; that is, the "OFF" position. The commonly used ON/OFF toggle switch is an SPST switch (single-pole, single-throw). A two-pole, single-throw switch opens or closes two conductors at the same time. Both conductors are either open or closed; that is, in the ON or OFF position. A two-pole, double-throw switch is used to direct a two-wire circuit through one of two different paths. One application of a two-pole, double-throw switch is in an electrical transfer switch where certain circuits may be energized from either the main electric service, or from an emergency standby generator. The double-throw switch "makes" the circuit from one or the other and prevents the circuits from being energized from both sources at once. Figure 8-3 shows common switch configurations.

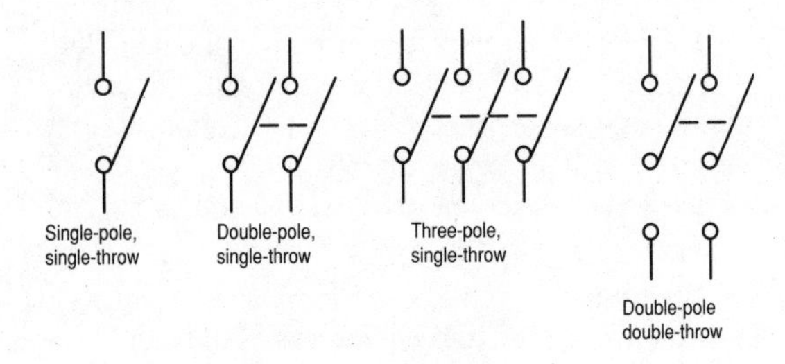

Figure 8-3: Common switch configurations.

Switch Identification

Switches vary in grade, capacity, and purpose. It is very important that proper types of switches are selected for the given application. For example, most single-pole toggle switches used for the control of lighting are restricted to ac use only. This same switch is not suitable for use on, say, a 32-V dc emergency lighting circuit. A switch rated for ac only will not extinguish a dc arc quickly enough. Not only is this a dangerous practice (causing arcing and heating of the device), the switch contacts would probably burn up after only a few operations of the handle, if not the first time.

Figure 8-4 shows a typical single-pole toggle switch — the type most often used to control ac lighting in all installations. Note the identifying marks.

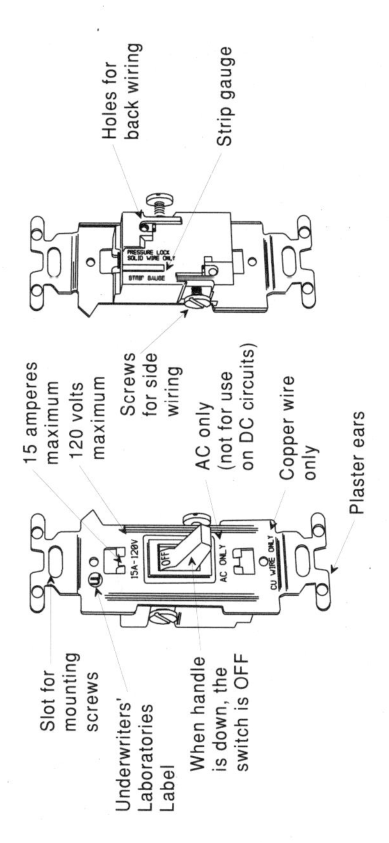

Figure 8-4: Typical identifying marks on a single-pole switch.

They are similar to those on the duplex receptacle discussed previously. The main difference is the "T" rating which means that the switch is rated for switching lamps with tungsten filaments (incandescent lamps).

Screw terminals are also color-coded on conventional toggle switches. Switches are typically constructed with a ground screw attached to the metallic strap of the switch. The ground screw is usually a green-colored hex-head screw. This screw is for connecting the equipment-grounding conductor to the switch. On three-way switches, the common or pivot terminal usually has a black or bronze screw head.

The switch shown is the type normally used for residential construction. Heavier-duty switches are usually the type used on commercial wiring — some of which are rated for use on 277-V circuits with current-carrying ratings up to 30 A. Therefore, it is important to check the rating of each switch before it is installed.

The exact type and grade of switch to be used on a specific installation is often dictated by the project drawings or written specifications. Sometimes wall switches are specified by manufacturer and catalog number; other times they are specified by type, grade, voltage, current rating, and the like, leaving the contractor or electrician to select the manufacturer. The naming of a certain brand of switch for a particular project, does not necessarily mean that this brand must be used.

Wiring diagrams of switch circuits — single-pole, three-way, and four-way switches — are discussed in Chapter 13 — Lighting Installations, while switches are covered in Chapter 9 — Panelboards and switchgear.

Chapter 9

PANELBOARDS AND SWITCHGEAR

Electricians usually become involved with electric service installations starting at the power company's point of attachment to the building. The connection is made from the power company's equipment to the building by either an overhead *service drop*, or an underground *service lateral*.

The remaining parts of an electric service are as follows:

- *Service entrance:* All components between the point of termination of the overhead service drop or underground service lateral and the building's main disconnecting device, except for metering equipment.
- *Service-entrance conductors:* The conductors between the point of termination of the overhead service drop or underground service lateral and the main disconnecting device in the building or on the premises.
- *Service-entrance equipment:* Provides overcurrent protection to the feeder and service conductors, a means of disconnecting the feeders from energized service conductors,

and a means of measuring the energy used by the use of metering equipment.

Common Power Supplies

The most common power supply for residential and small commercial applications is the 120/240 V, single-phase service; it is used primarily for light and power, including single-phase motors up to about 7½ horsepower (hp). A diagram of this type of service is shown in Figure 9-1.

Four-wire wye-connected secondaries (Figure 9-2) and four-wire, delta-connected secondaries (Figure 9-3) are common around larger commercial and industrial facilities.

The characteristics of the electric service and the connected equipment must match; also, the characteristics of an electric service will often dictate those for the electrical equipment or vice versa.

Service Disconnecting Means

A service disconnecting means is a device or devices that enable the electric service to be disconnected from the building premises. Several different configurations are possible.

Service switches, load centers, or main distribution panelboards are normally installed at a point im-

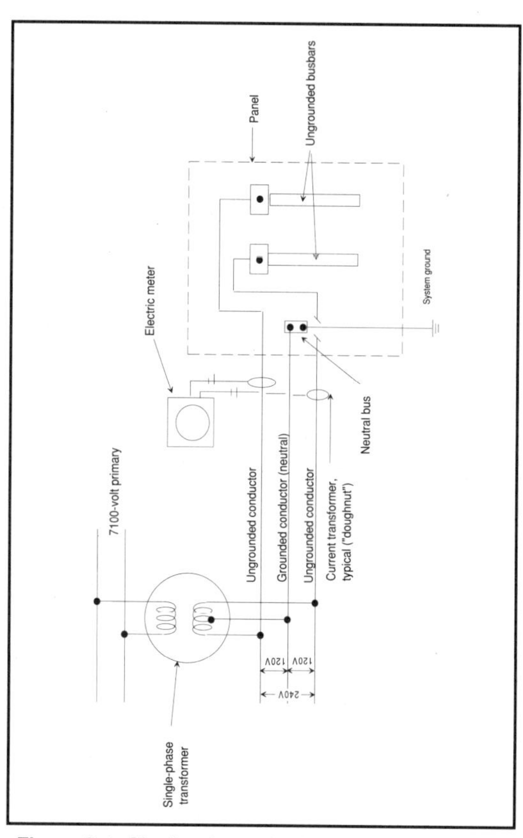

Figure 9-1: Single-phase, 3-wire, 120/240-V electric service.

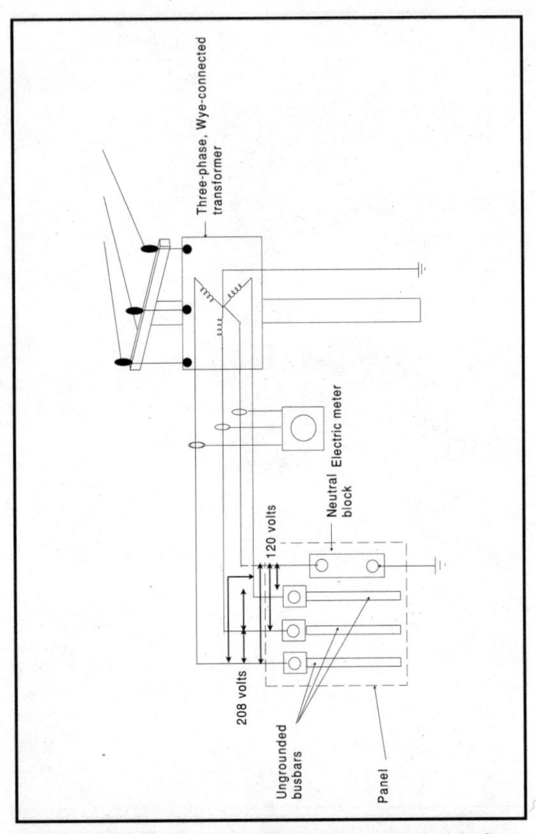

Figure 9-2: Three-phase, 4-wire, wye-connected electric service.

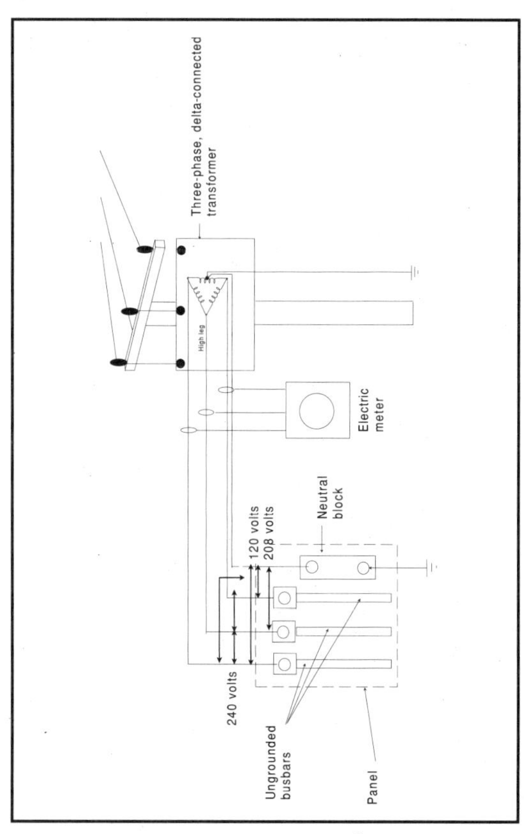

Figure 9-3: Three-phase, 4-wire, delta-connected electric service.

mediately where the service conductors enter the building. Branch circuits and feeder panelboards are usually grouped together at one or more centralized locations to keep the length of the branch-circuit conductors at a practical minimum of operating efficiency and to lower the initial installation costs.

Distribution panelboards or equipment is generally intended to carry and control electrical current, but is not intended to dissipate or utilize energy. Eight basic factors influence the selection of distribution equipment:

1. *Codes and Standards*: Suitability for installation and use, in conformity with the provisions of the *NEC* and all local codes, must be considered. Suitability of equipment may be evidenced by listing or labeling.

2. *Mechanical Protection*: Mechanical strength and durability, including the adequacy of the protection provided must be considered.

3. *Wiring Space*: Wire bending and connection space is provided according to UL standards in all distribution equipment. When unusual wire arrangements or connections are to be

made, then extra wire bending space, gutters, and terminal cabinets should be investigated for use.

4. *Electrical Insulation*: All distribution equipment carries labels showing the maximum voltage level that should be applied. The electrical supply voltage should always be equal to, or less than, the voltage rating of distribution equipment; never more.

5. *Heat*: Heating effects under normal conditions of use and also under abnormal conditions likely to arise in service must be constantly considered. Ambient heat conditions, as well as wire insulation ratings, along with the heat rise of the equipment must be evaluated during selection.

6. *Arcing Effects*: The normal arcing effects of overcurrent protective devices must be considered when the application is in or near combustible materials or vapors. Enclosures are selected to prevent or contain fires created by normal operation of the equipment. Selected locations of

equipment must be made when another location may cause a hazardous condition.

7. *Classification*: Classification according to type, size, voltage, current capacity, interrupting capacity and specific use must be considered when selecting distribution equipment. Loads may be continuous or noncontinuous and the demand factor must be determined before distribution equipment can be selected.

8. *Personal Protection*: Other factors that contribute to the practical safeguarding of a person using or likely to come in contact with the equipment must be considered. The equipment selected for use by only qualified persons may be different from equipment used or applied where unqualified people may come in contact with it.

In electrical wiring installations, overcurrent protective devices, consisting of fuses or circuit breakers, are sometimes factory assembled in a metal cabinet, the entire assembly commonly being called a *panelboard*. At other times, the panelboards will be delivered unassembled, consisting of an enclosure ("can"), the interior busbars, and the trim. Circuit breakers are then installed as the project dictates.

Sometimes the main service-disconnecting means will be made up on the job by assembling individually enclosed fused switches or circuit breakers on a length of metal auxiliary gutter.

Grounding

NEC Article 250 covers general requirements for grounding and bonding electric services. In general, the *NEC* requires a premises wiring system, supplied by an alternating-current service to be grounded by a grounding electrode conductor connected to a grounding electrode. The grounding electrode conductor must be bonded to the grounded service conductor (neutral) at any accessible point from the load end of the service drop or service lateral to, and including, the terminal bus to which the grounded service conductor is connected at the service disconnecting means. A grounding connection must not be made to any grounded circuit conductor on the load side of the service disconnecting means.

Most applications require the grounded service conductor to be bonded to at least two grounding electrodes according to *NEC* Section 250-81.

The table in Figure 9-4 gives the required sizes of grounding conductors for various sizes of electric services.

Size of Largest Service-Entrance Conductor or Equivalent for Parallel Conductors		Size of Grounding Electrode Conductor	
Copper	**Aluminum or Copper-Clad Aluminum**	**Copper**	**Aluminum or Copper-Clad Aluminum**
2 or smaller	0 or smaller	**8**	**6**
1 or 2	2/0 or 3/0	**6**	**4**
2/0 or 3/0	4/0 or 250 kcmil	**4**	**2**
Over 3/0 through 350 kcmil	Over 250 kcmil through 500 kcmil	**2**	**0**
Over 350 kcmil through 600 kcmil	Over 500 kcmil through 900 kcmil	**0**	**3/0**
Over 600 kcmil through 1100 kcmil	Over 900 kcmil through 1750 kcmil	**2/0**	**4/0**
Over 1100 kcmil	Over 1750 kcmil	**3/0**	**250 kcmil**

Figure 9-4: Minimum sizes of grounding electrode conductors for ac system.

Switches, Panelboards, and Load Centers

Panelboards consist of assemblies of overcurrent protective devices, with or without disconnecting devices, placed in a metal cabinet. The cabinet includes a cover or trim with one or two doors to allow access to the overcurrent and disconnecting devices and, in some types, access to the wiring space in the panelboard.

Panelboards fall into two mounting classifications:

- Flush mounting (Figure 9-5)
- Surface mounting (Figure 9-6)

When panelboards and cabinets are flush mounted, the trim extends beyond the outside edges of the cabinet to provide a neat finish with the wall. When surface mounted, the trim is flush with the edge of the cabinet.

Panelboards also fall into two general classifications with regard to overcurrent protective devices:

- Circuit breakers
- Fuses

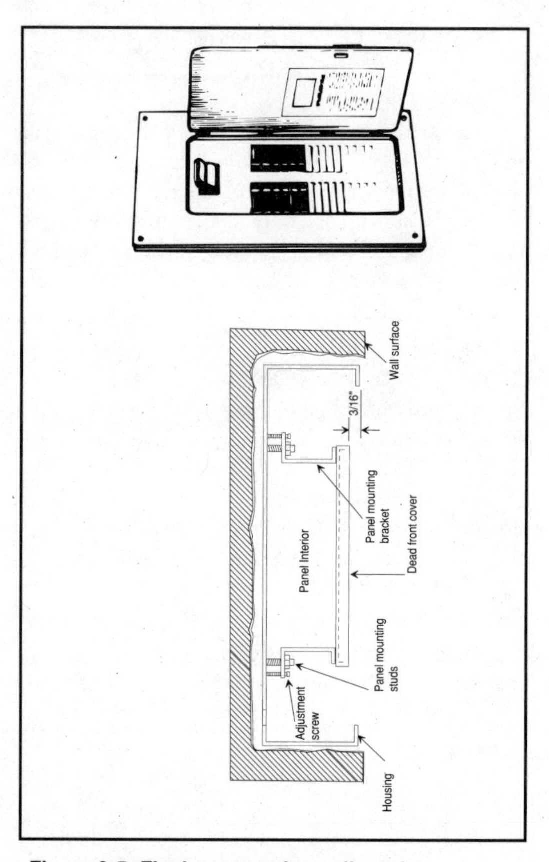

Figure 9-5: Flush-mounted panelboard.

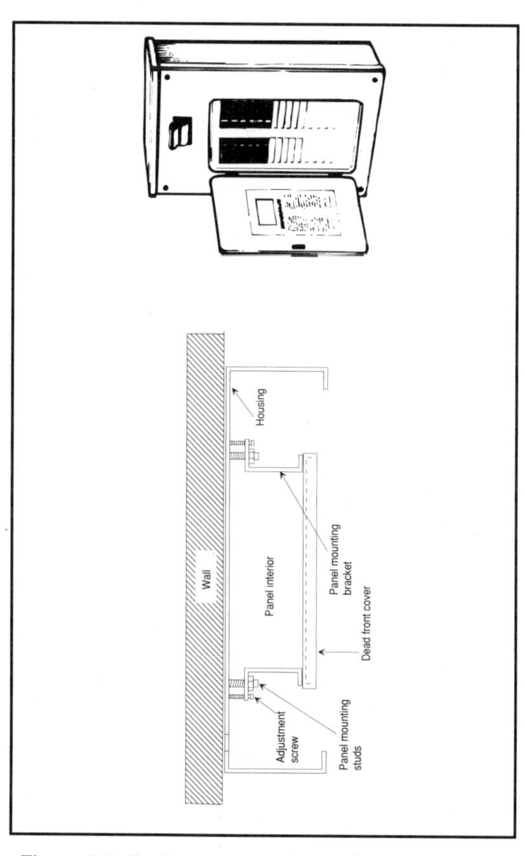

Figure 9-6: Surface-mounted panelboard.

Knockouts

A series of concentric or eccentric circular partial openings are usually cut in the top, bottom, and sides of both load centers and panelboard housings. These openings are cut in such a manner that they may be removed by tapping (knocking) them out — usually with a screwdriver blade and hammer. See Figure 9-7.

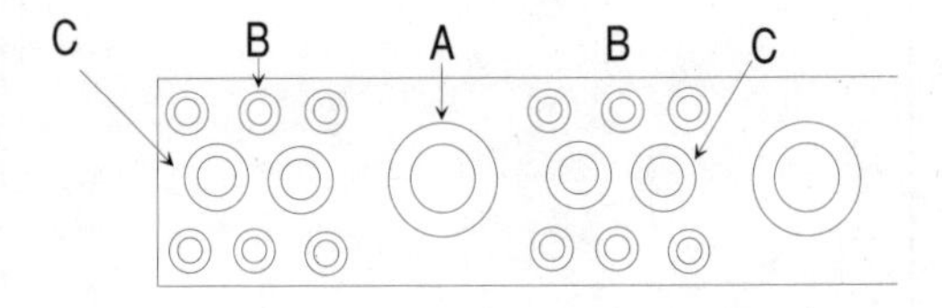

A	Two 2- or 2½-inch knockouts; space for two 3½-inch knockouts
B	Twelve ½- or ¾-inch knockouts
C	Four ¾- or 1-inch knockouts

Figure 9-7: Typical knockouts in panelboard top.

Panel Connections

Electrical connections in a panelboard fall under two categories:

- Line connections, which include termination and routing of the service and feeder conductors.
- Load connections, which include termination and routing of the branch-circuit and feeder conductors.

When installing the line connections, verify that the lugs are stamped "CU/AL" or a label is inside the panel which states that the connection of aluminum conductors is permitted prior to terminating aluminum conductors.

When installing the load connections, again verify that the lugs are suitable for both copper and aluminum if aluminum conductors are used. Due to the difficulty in keeping aluminum conductors tight in their termination lugs, copper is usually specified for most industrial installations.

NEC Section 384-3 requires the phase arrangement on 3-phase buses to be A, B, C from front to back, top to bottom, or left to right, as viewed from the front of the panel. The B phase must be the phase with the highest voltage to ground on a 3-phase, 4-wire delta-connected system.

Enclosures

The majority of overcurrent devices (fuses and circuit breakers) are used in some type of enclosure; that is, panelboards, switchboards, motor-control centers, individual enclosures, etc.

NEMA has established enclosure designations because individually enclosed overcurrent-protective devices are used in so many different types of locations, weather and water conditions, dust and other contaminating conditions, etc. A designation such as "NEMA 12" indicates an enclosure type to fulfill requirements for a particular application. The NEMA designations were recently revised to obtain a clearer and more precise definition of the enclosure needed to meet various standard requirements.

Some of the revisions in the NEMA designations are: The NEMA Type 1A (semi-dust tight) has been dropped. The NEMA 12 enclosure now can be substituted in many installations in place of the NEMA 5. The advantage of this substitution is that the NEMA 12 enclosure is much less expensive than the NEMA 5 enclosure. NEMA Type 3R as applied to circuit-breaker enclosures is a lighter weight, less expensive rainproof enclosure than the other "Weather Resistant" enclosure types.

Safety Switches

Most manufacturers of safety switches have at least two complete lines to meet industrial, commercial and residential requirements. Both types usually have visible blades and safety handles. With visible blades, the contact blades are in full view so you can clearly see you're safe. Safety handles are always in complete control of the switch blades, so whether the cover is open or closed, when the handle is in the "OFF" position the switch is always "OFF"; that is, on the load side of the switch. The feeder or line side of the switch is still "hot" (energized) so when working with safety switches keep this in mind.

WARNING!

Even though a safety-switch handle is in the OFF position, the line side of the switch is still energized.

Heavy duty switches are intended for applications where price is secondary to safety and continued performance. This type of switch is usually subjected to frequent operation and rough handling. Heavy duty switches are also used in atmospheres where a general-duty switch would be unsuitable. Heavy-duty switches are widely used by most heavy industrial applications; motors and HVAC equipment will also be controlled by such switches. Most heavy-duty switches are rated 30 through 1200 amperes, 240 to 600 volts (ac-dc). The switches with horsepower ratings are able to interrupt approximately six times the full-load, motor-current ratings. When equipped with Class J or Class R fuses , many heavy-duty safety switches are UL listed for use on systems with up to 200,000 A available fault current.

Heavy-duty switches are available with NEMA 1, 3R, 4, 4X, 5, 7, 9 and 12 enclosures.

Switch Contacts: There are two types of switch contacts used in today's safety switches. One is the "butt" contact similar to those used in circuit-breaker devices; the other is a knife-blade and jaw type. The knife-blade types are considered to be superior to other types on the market.

All current-carrying parts of safety switches are usually plated with tin, cadmium or nickel to reduce heating by keeping metal oxidation at a minimum. Switch blade and jaws are made of copper for high conductivity. With knife-blade construction, the jaws distribute a uniform clamping pressure over the entire blade-to-jaw contact surface. In the event of high-current fault, the electromagnetic forces which develop tend to squeeze the jaws tightly against the blade. In the butt-type contact, these forces tend to force the contacts apart, causing them to burn severely.

Fuse clips are also plated to control corrosion and keep heating to a minimum. All heavy-duty fuse clips have steel reinforcing springs to increase their mechanical strength and give a firmer contact pressure. As a result, fuses will not work loose due to vibration or rough handling.

Insulating Materials: As the voltage rating of switches is increased, arc suppression becomes more difficult and the choice of insulation material becomes a more critical problem. Arc suppressors used by many manufacturers consist of a housing made of insulation material and one or more magnetic suppressor plates. All arc suppressors are tested to assure proper control and extinguishing of arcing.

Operating Mechanism: Heavy-duty safety switches have spring-driven, quick-make, quick-break mechanisms. A quick-breaking action is necessary if a switch is to be safely switched "OFF" under a heavy load. The spring action, in addition to making the operation quick-make, quick-break, firmly holds the switch blades in an "ON" or "OFF" position. The operating handle is an integral part of the switching mechanism, so if the springs should fail the switch can still be operated. When the handle is in the "OFF" position the switch is always "OFF."

A one-piece cross bar is usually employed to offer direct control over all blades simultaneously.

General-Duty Safety Switches: General-duty switches are for residential and light commercial applications where the price of the device is a limiting factor. General duty switches are meant to be used where operation and handling are moderate and where the available fault current is less than 10,000 A. Some examples of general duty switch applications would be: residential HVAC equipment, light duty fan-coil circuit disconnects for commercial projects, and the like.

General-duty switches are rated up to 600 amperes at 240 volts (ac only) in general purpose (NEMA 1) and rainproof (NEMA 3R) enclosures. These switches are horsepower rated and capable of opening a circuit with approximately six times a motor's full-load current rating.

All current-carrying parts of general-duty switches are plated with either tin or cadmium to reduce heating. Switch jaws and blades are made of plated copper for high conductivity. A steel reinforcing spring increases the mechanical strength of the jaws and assures a firm contact pressure between blade and jaw.

Double-Throw Safety Switches: Double-throw switches are used as transfer switches and are not intended as motor circuit switches; therefore, they are not horsepower rated.

Chapter 10

CONDUCTOR TERMINATIONS

Anyone involved with electrical systems of any type should have a good knowledge of wire connectors and splicing, as it is necessary to make numerous electrical joints during the course of any electrical installation.

Splices and connections that are properly made will often last as long as the insulation on the wire itself, while poorly made connections will always be a source of trouble; that is, the joints will overheat under load and cause a higher resistance in the circuit than there should be.

The basic requirements for a good electrical connection include the following:

- It should be mechanically and electrically secure.
- It should be insulated as well as, or better than, the existing insulation on the conductors.
- These characteristics should last as long as the conductor is in service.

There are many different types of electrical joints for different purposes, and the selection of the proper type for a given application will depend to a great extent on how and where the splice or connection is used.

Electrical joints are normally made with solderless pressure connectors or lugs to save time, but electricians should also have a knowledge of the traditional splices. Most traditional splices appear in Figure 10-1.

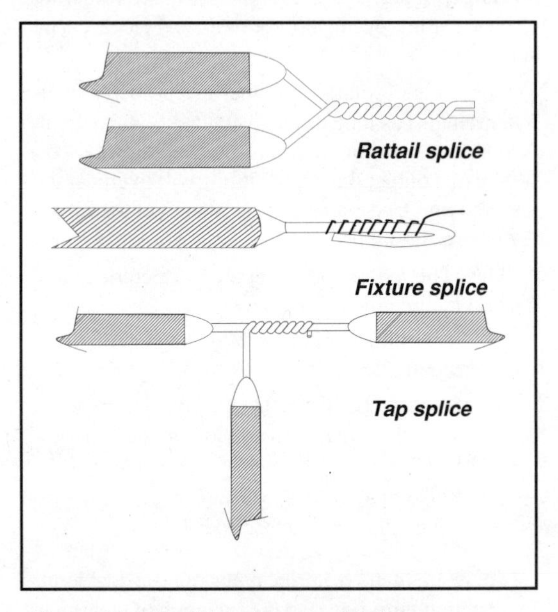

Figure 10-1: Traditional taps and splices.

Stripping and Cleaning

Before any connection or splice can be made, the ends of the conductors must be properly stripped and cleaned. Stripping is the removal of insulation from the conductors at the end of the wire or at the location of the splice. Some electricians strip the smaller sizes of wire with a pocket knife or a pair of side-cutting pliers, but there are many handy tools on the market that will facilitate this operation. The use of such tools will also help to prevent cuts and nicks in the wire which reduce the conductor area as well as weaken it.

Poorly stripped wire can result in nicks, scrapes, or burnishes. Any of these can lead to a stress concentration at the damaged cross section. Heat, rapid temperature change, mechanical vibration, and oscillatory motion can aggravate the damage, causing faults in the circuitry or even total failure.

Lost strands are a problem in splices or crimp-type terminals, while exposed strands might be a safety hazard.

Slight burnishes on conductors, as long as they had no sharp edges, were acceptable at one time. Now, however, reliable experts feel that under certain conditions removing as little as 40 micro inches of conductor plating from some wires can cause failure.

Faulty stripping can pierce, scuff, or split the insulation. This can cause changes in dielectric strength and lower the wire's resistance to moisture and abrasion. Insulation particles often get trapped in solder and crimp joints. These form the basis for a defective termination. A variety of factors determine just how precisely a wire can be stripped: wire size, insulation concentricity, adherence, and others.

It is a common mistake to believe that a certain gauge of stranded conductor has the same diameter as a solid conductor. Stranded conductors of the same wire-gauge size are always larger in diameter than solid conductors of the same gauge. This is a very important consideration in selecting proper blades for wire strippers. The table in Figure 10-2 shows the nominal sizes referenced for the different wire gauges.

Wire Connections Under 600 V

Wire connections are used to connect a wire or cable to such electrically-operated devices as fan-coil units, duct heaters, oil burners, motors, pumps, and control circuits of all types.

A variety of wire connectors are shown in Figure 10-3. These connectors are available in various sizes to accommodate wire from No. 22 AWG through 250 kcmil. They can be installed with crimping tools

DIMENSIONS OF COMMON WIRE SIZES			
Size, AWG/kcmil	Area Circular Mils	Overall Diameter in Inches	
		Solid	Stranded
18	1620	0.040	0.046
16	2580	0.051	0.058
14	4130	0.064	0.073
12	6530	0.081	0.092
10	10380	0.102	0.116
8	16510	0.128	0.146
6	26240	-	0.184
4	41740	-	0.232

Figure 10-2: Dimensions of common wire sizes.

DIMENSIONS OF COMMON WIRE SIZES *(Cont.)*			
Size, AWG/kcmil	Area Circular Mils	Overall Diameter in Inches	
		Solid	Stranded
3	52620	-	0.260
2	66360	-	0.292
1	83690	-	0.332
1/0	105600	-	0.373
2/0	133100	-	0.419
3/0	167800	-	0.470
4/0	211,600	—	0.528
250	250,000	—	0.575

Figure 10-2: Dimensions of common wire sizes. *(Cont.)*

DIMENSIONS OF COMMON WIRE SIZES *(Cont.)*			
Size, AWG/kcmil	Area Circular Mils	Overall Diameter in Inches	
		Solid	Stranded
300	300,000	—	0.630
350	350,000	—	0.681
400	400,00	—	0.728
500	500,000	—	0.813
600	600,000	—	0.893
700	700,000	—	0.964
750	750,000	—	0.998
800	800,000	—	1.03

Figure 10-2: Dimensions of common wire sizes. ***(Cont.)***

DIMENSIONS OF COMMON WIRE SIZES *(Cont.)*			
Size, AWG/kcmil	**Area Circular Mils**	**Overall Diameter in Inches**	
		Solid	**Stranded**
900	900,000	—	1.09
1000	1,000,000	—	1.15
1250	1,250,000	—	1.29
1500	1,500,000	—	1.41
1750	1,750,000	—	1.52
2000	2,000,000	—	1.63

Figure 10-2: Dimensions of common wire sizes. ***(Cont.)***

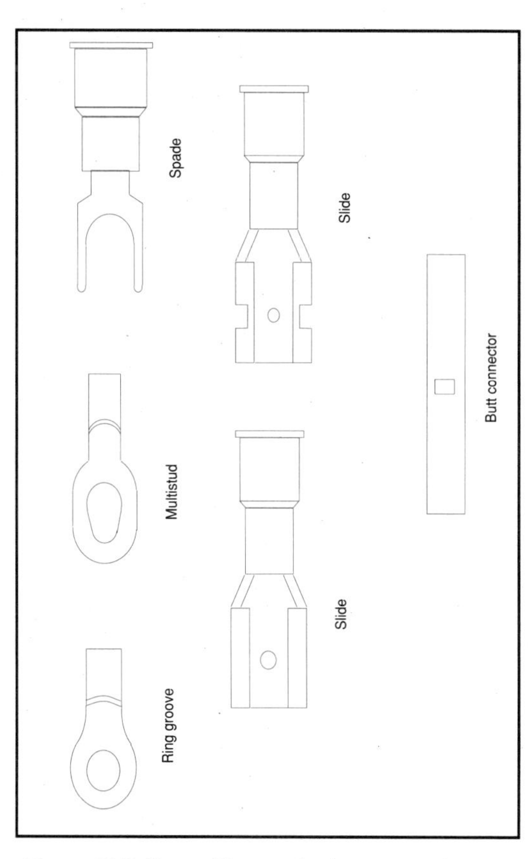

Figure 10-3: Several types of crimp connectors.

having a single indenter or double indenter. Wide range identification is normally stamped on the tongue of each terminal.

Compression type terminators are also available to accommodate wires from No. 8 AWG through 1000 kcmil. One-hole lugs, two-hole lugs, and split-bolt connectors are shown in Figure 10-4.

Aluminum Connections

Aluminum has certain properties that are different from copper that must be understood if reliable connections are to be made. These properties are: cold flow, coefficient of thermal expansion, susceptibility to galvanic corrosion, and the formation of oxide film on the metal's surface.

Because of thermal expansion and cold flow of aluminum, standard copper connectors as found on the market today cannot be safely used on aluminum wire. Most manufacturers design their aluminum connectors with greater contact area to counteract this property of aluminum. Tongues and barrels of all aluminum connectors are larger or deeper than comparable copper connectors.

The electrolytic action between aluminum and copper can be controlled by plating the aluminum with a neutral metal (usually tin). The plating prevents electrolysis from taking place and the joint

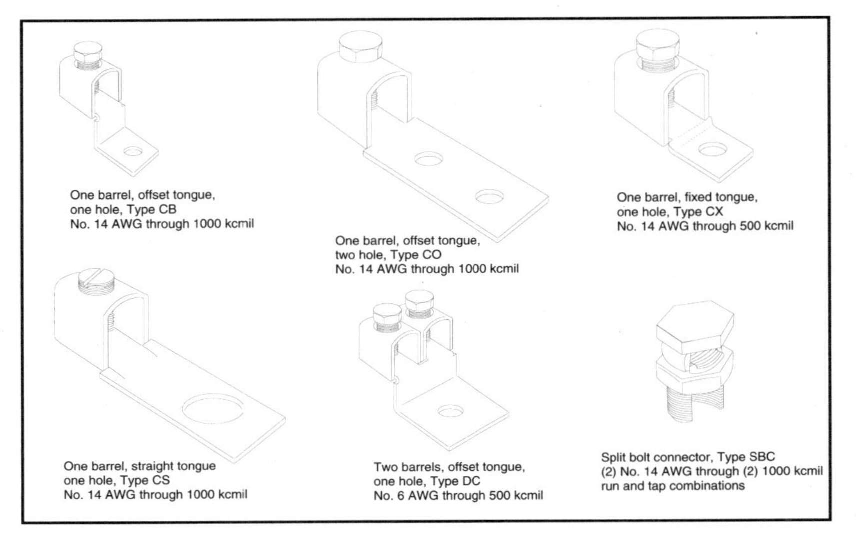

Figure 10-4: Types of compression connectors.

remains tight. As an additional precaution, a joint sealing compound should be used. Connectors should be tin plated and prefilled with an oxide abrading sealing compound.

The insulating aluminum oxide film must be removed or penetrated before a reliable aluminum joint can be made. Aluminum connectors are designed to bite through this film as they are applied to conductors. It is further recommended that the conductor be wire brushed and preferably coated with a joint compound to guarantee a reliable joint.

- Connectors marked with just the wire size should only be used with copper conductors.
- Connectors marked with Al and the wire size should only be used with aluminum wire.
- Connectors marked with Al-Cu and the wire size may be safely used with either copper or aluminum.

Wire Nuts

Ever since its invention in 1927, the wire nut (Figure 10-5) has been a favorite wire connector for use on branch-circuit applications where permitted by the *NEC*. Several varieties of wire nuts are available, but the following are the ones used the most:

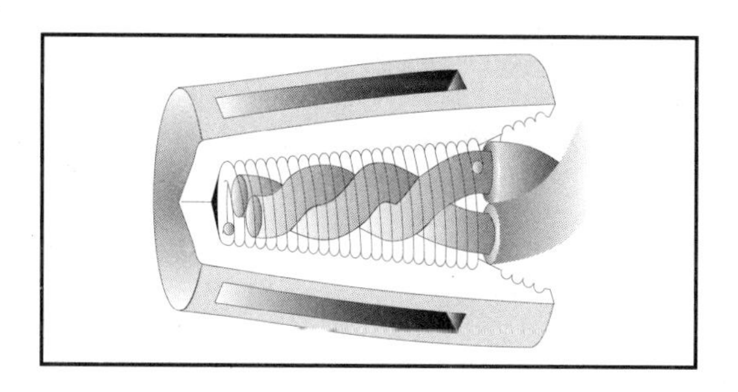

Figure 10-5: Typical wire nut showing interior arrangement.

- For use on wiring systems 300 V and under.
- For use on wiring systems 600 V and under (1000 V in lighting fixtures and signs).

Most brands are UL listed for aluminum to copper in dry locations only; aluminum to aluminum only; or copper to copper only; 600 V maximum; 1000 V in lighting fixtures and neon signs. The maximum temperature rating is 105°C (221°F).

Wire nuts are frequently used for all types of splices in residential and commercial applications and are considered to be the fastest connectors on the market for this type of work.

Traditionally, electricians form a pigtail splice on the ends of conductors with side-cutting pliers, trim

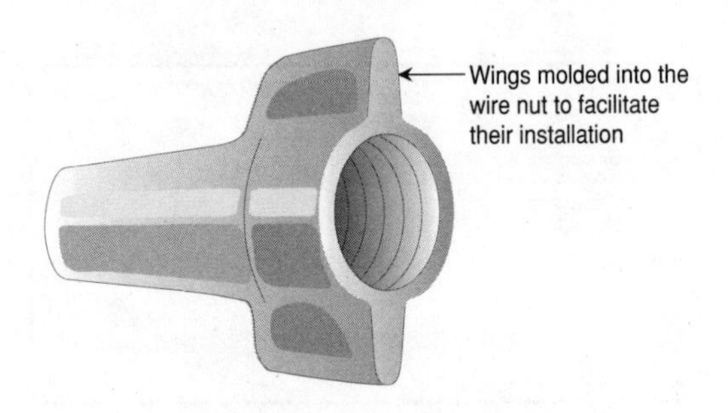

Figure 10-6: Some types of wire nuts have thin wings on each side to facilitate installation.

the bare conductors with the pliers' cutters, and then screw on the wire nut. In doing so, the wire nut draws the conductors and insulation into the shirt of the connector which increases resistance to flashover. The internal spring is designed to tightly "thread" the conductors into the wire nut and then hold them with a positive grip. Some type of wire nuts have thin wings on each side of the connector to facilitate their installation. *See* Figure 10-6.

Wire nuts are made in sizes to accommodate conductors as small as No. 22 AWG up to as large as No. 10 AWG, with practically any combination of those sizes in between.

The model numbers of wire nuts will vary with the manufacturer, but Ideal set a standard years ago with their 71B through 76B series.

Chapter 11

ELECTRIC MOTORS

Electric motors have long been the workhorses of practically every kind of installation from residential appliances to heavy industrial machines. Many types of motors are available from small shaded-pole motors (used mostly in household fans) to huge synchronous motors for use in large industrial installations. There are several types in between to fill every conceivable niche. None, however, have the wide application possibilities of the three-phase motor. This is the type of motor that electricians will work with the most. Therefore, the majority of the material in this chapter will deal with three-phase motors.

There are three basic types of three-phase motors:

- The squirrel cage induction motor.
- The wound rotor induction motor.
- The synchronous motor.

The type of three-phase motor is determined by the rotor or rotating member (*see* Figure 11-1). The stator winding for any of these motors is the same.

The principle of operation for all three-phase motors is the rotating magnetic field. There are three factors that cause the magnetic field to rotate:

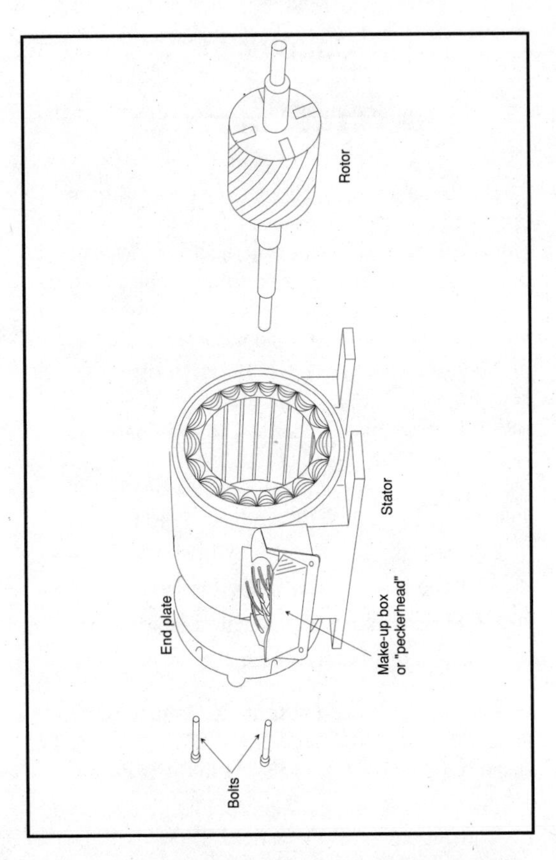

Figure 11-1: Basic parts of a three-phase motor.

- The voltages of a three-phase electrical system are 120° out of phase with each other.
- The three voltages change polarity at regular intervals.
- The arrangement of the stator windings around the inside of the motor.

The *NEC* also plays an important role in the installation of electric motors. *NEC* Article 430 covers application and installation of motor circuits and motor control connections — including conductors, short-circuit and ground-fault protection, controllers, disconnects, and overload protection.

NEC Article 440 contains provisions for motor-driven air conditioning and refrigerating equipment — including the branch circuits and controllers for the equipment. It also takes into account the special considerations involved with sealed (hermetic-type) motor compressors, in which the motor operates under the cooling effect of the refrigeration. In referring to *NEC* Article 440, be aware that the rules in this *NEC* Article are *in addition to*, or are *amendments to*, the rules given in *NEC* Article 430.

Motors are also covered to some degree in *NEC* Articles 422 and 424.

Synchronous Speed

The speed at which the magnetic field rotates is known as the synchronous speed. The synchronous speed of three-phase motors is determined by two factors:

- Number of stator poles
- Frequency of the ac line

Since 60 Hz is the standard frequency throughout the United States and Canada, the following gives the synchronous speeds for motors with different numbers of poles:

- 2 poles — 3600 rpm
- 4 poles — 1800 rpm
- 6 poles — 1200 rpm
- 8 poles — 900 rpm

Motor Connections

The stator windings of three-phase motors are connected in either wye or delta. Some motors are designed to operate both ways; that is, started as a wye-connected motor to help reduce starting current, and then changed to a delta connection for running. See the following illustrations for various motor connections.

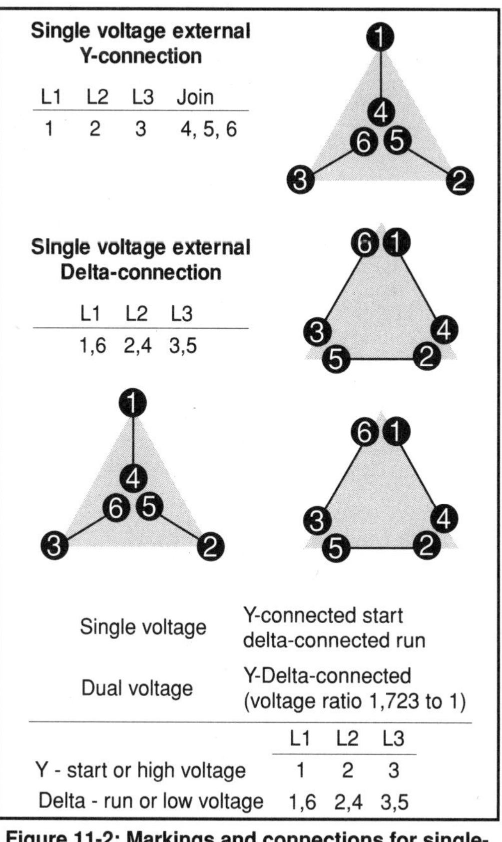

Figure 11-2: Markings and connections for single-speed three-phase motors with six leads.

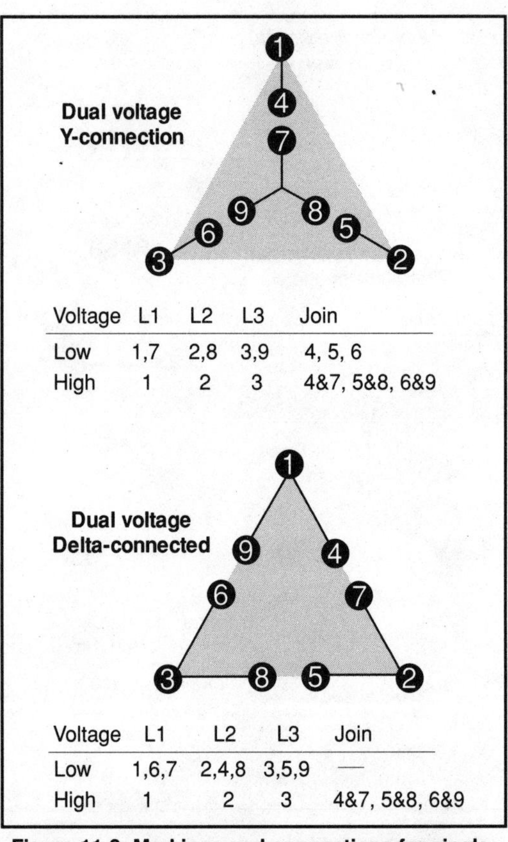

Voltage	L1	L2	L3	Join
Low	1,7	2,8	3,9	4, 5, 6
High	1	2	3	4&7, 5&8, 6&9

Voltage	L1	L2	L3	Join
Low	1,6,7	2,4,8	3,5,9	—
High	1	2	3	4&7, 5&8, 6&9

Figure 11-3: Markings and connections for single-speed three-phase motors with nine leads.

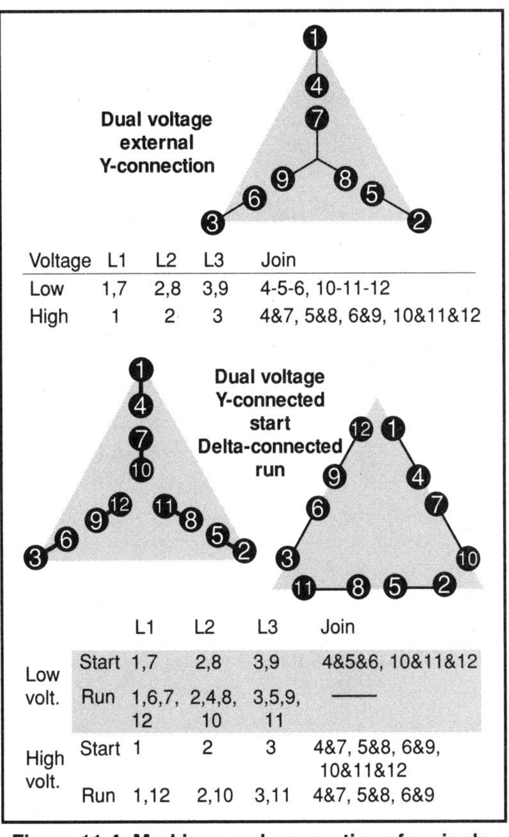

Voltage	L1	L2	L3	Join
Low	1,7	2,8	3,9	4-5-6, 10-11-12
High	1	2	3	4&7, 5&8, 6&9, 10&11&12

		L1	L2	L3	Join
Low volt.	Start	1,7	2,8	3,9	4&5&6, 10&11&12
	Run	1,6,7, 12	2,4,8, 10	3,5,9, 11	——
High volt.	Start	1	2	3	4&7, 5&8, 6&9, 10&11&12
	Run	1,12	2,10	3,11	4&7, 5&8, 6&9

Figure 11-4: Markings and connections for single-speed three-phase motors with 12 leads.

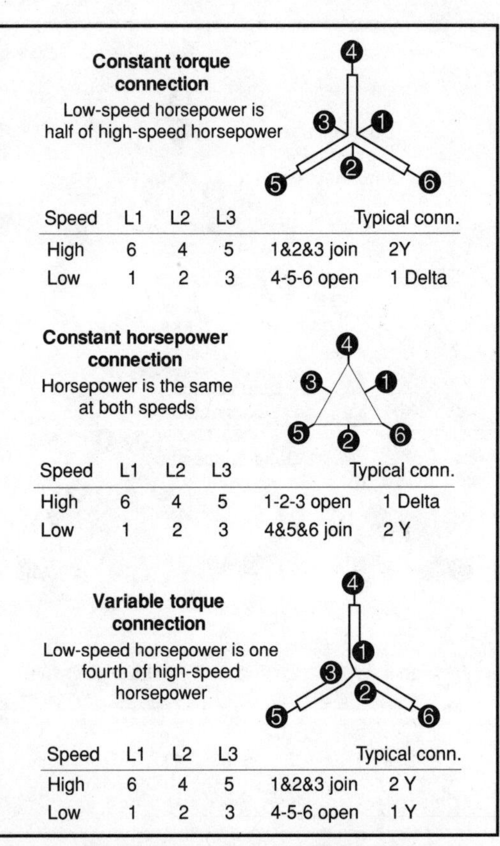

Speed	L1	L2	L3		Typical conn.
High	6	4	5	1&2&3 join	2Y
Low	1	2	3	4-5-6 open	1 Delta

Speed	L1	L2	L3		Typical conn.
High	6	4	5	1-2-3 open	1 Delta
Low	1	2	3	4&5&6 join	2 Y

Speed	L1	L2	L3		Typical conn.
High	6	4	5	1&2&3 join	2 Y
Low	1	2	3	4-5-6 open	1 Y

Figure 11-5: Markings and connections for two-speed three-phase motors with 6 leads.

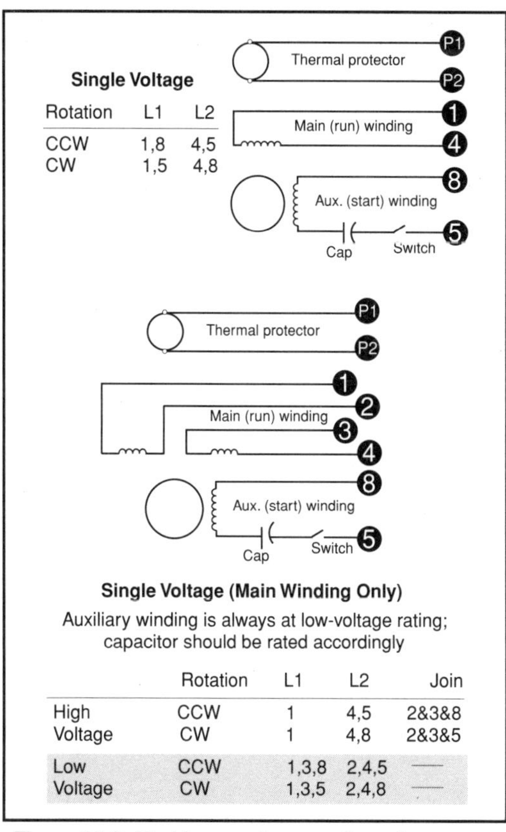

Rotation	L1	L2
CCW	1,8	4,5
CW	1,5	4,8

	Rotation	L1	L2	Join
High Voltage	CCW	1	4,5	2&3&8
	CW	1	4,8	2&3&5
Low Voltage	CCW	1,3,8	2,4,5	—
	CW	1,3,5	2,4,8	—

Figure 11-6: Markings and connections for capacitor-start single-phase motors.

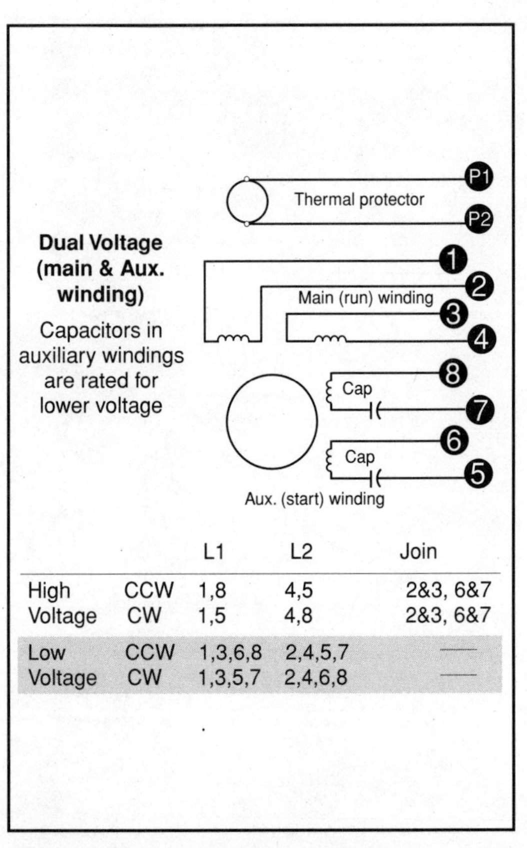

		L1	L2	Join
High Voltage	CCW	1,8	4,5	2&3, 6&7
	CW	1,5	4,8	2&3, 6&7
Low Voltage	CCW	1,3,6,8	2,4,5,7	—
	CW	1,3,5,7	2,4,6,8	—

Figure 11-7: Markings and connections for Capacitor-start single-phase motors.

Chapter 12

MOTOR CONTROLS

Electric motors provide one of the principal sources for driving all types of equipment and machinery. Every motor in use, however, must be controlled, if only to start and stop it, before it becomes of any value.

Motor controllers cover a wide range of types and sizes, from a simple toggle switch to a complex system with such components as relays, timers, and switches. The common function, however, is the same in any case, that is, to control some operation of an electric motor. A motor controller will include some or all of the following functions:

- Starting and stopping
- Overload protection
- Overcurrent protection
- Reversing
- Changing speed
- Jogging
- Plugging
- Sequence control
- Pilot light indication

The controller can also provide the control for auxiliary equipment such as brakes, clutches, solenoids, heaters, and signals, and may be used to control a single motor or a group of motors.

The term *motor starter* is often used and means practically the same thing as a *controller*. Strictly, a motor starter is the simplest form of controller and is capable of starting and stopping the motor and providing it with overload protection.

Motor Nameplates

Much information about sizing motor starters can be found from the motor's nameplate. Consequently, a review of motor nameplates is in order.

A typical motor nameplate is shown in Figure 12-1. A nameplate is one of the most important parts of a motor since it gives the motor's electrical and mechanical characteristics; that is, the horsepower, voltage, rpms, etc. Always refer to the motor's nameplate before connecting it to an electric system or selecting the motor starter and related components. The same is true when performing preventative maintenance or troubleshooting motors.

Referring again to the motor nameplate in Figure 12-1, note that the manufacturer's name and logo is at the top of the plate; these items, of course, will change with each manufacturer. The line directly below the

TURNER		ELECTRIC			
A.C. MOTOR					
MODEL	674	TYPE	KC	PHASE	1
HP	$1\frac{1}{2}$	HERTZ (CYCLES)	60	F.L.A.	15/7.5
VOLTS	120/240	RPM	1725		
TEMP. RISE	40° C	L.R.A.	80/40		
DUTY RATING	CONTINUOUS			SERIAL NO.	018204
CODE	D88	S.F.	1.25	F.R.	66

Figure 12-1: Typical motor nameplate.

manufacturer's name identifies the motor for use on ac systems as opposed to dc or ac-dc systems. The model number identifies that particular motor from any other. The type or class specifies the insulation used to ensure the motor will perform at the rated horsepower and service-factor load. The phase indicates whether the motor has been designed for single- or three-phase use.

When selecting motor starters for a given motor, the motor code letter on the nameplate plays an important role. The chart in Figure 12-2 may be used as a guide when selecting motor controllers.

Overload Protection

Overload protection for an electric motor is necessary to prevent burnout and to ensure maximum operating life. Electric motors will, if permitted, operate at

NEMA SIZE	Volts	Maximum HP Nonplugging and Nonjogging Duty		Maximum HP Rating Plugging and Jogging Duty	
		Single Phase	Polyphase	Single Phase	Polyphase
00	115	⅓	—	—	—
	200	—	1½	—	—
	230	1	1½	—	—
	380	—	1½	—	—
	460	—	2	—	—
	575	—	2	—	—
0	115	1	—	½	—
	200	—	3	—	1½
	230	2	3	1	1½
	380	—	5	—	1½
	460	—	5	—	2
	575	...	5	...	2

Figure 12-2: Electrical ratings for ac magnetic contactors and starters.

NEMA SIZE	Volts	Maximum HP Nonplugging and Nonjogging Duty		Maximum HP Rating Plugging and Jogging Duty	
1	115	2	—	1	—
	200	—	7½	—	3
	230	3	7½	2	3
	380	—	10	—	5
	460	—	10	—	5
	575	—	10	—	5
1P	115	3	—	1½	—
	230	5	—	3	—
2	115	3	—	2	—
	200	—	10	—	7½
	230	7½	15	5	10

Figure 12-2: Electrical ratings for ac magnetic contactors and starters. ***(Cont.)***

NEMA SIZE	Volts	Maximum HP Nonplugging and Nonjogging Duty		Maximum HP Rating Plugging and Jogging Duty	
2	380	—	25	—	15
	460	—	25	—	15
	575	—	25	—	15
3	115	7½	—	—	—
	200	—	25	—	15
	230	15	30	—	20
	380	—	50	—	30
	460	—	50	—	30
	575	—	50	—	30

NEMA SIZE	Volts	Maximum HP Nonplugging and Nonjogging Duty		Maximum HP Rating Plugging and Jogging Duty	
4	200	—	40	—	25
	230	—	50	—	30
	380	—	75	—	50
	460	—	100	—	60
	575	—	100	—	60
5	200	—	75	—	60
	230	—	100	—	75
	380	—	150	—	125
	460	—	200	—	150
	575	—	200	—	150
6	200	—	150	—	125
	230	—	200	—	150

Figure 12-2: Electrical ratings for ac magnetic contactors and starters. ***(Cont.)***

an output of more than rated capacity. Conditions of motor overload may be caused by an overload on driven machinery, by a low line voltage, or by an open line in a polyphase system, which results in single-phase operation. Under any condition of over-

NEMA SIZE	Volts	Maximum HP Nonplugging and Nonjogging Duty		Maximum HP Rating Plugging and Jogging Duty	
6	380	—	300	—	250
	460	—	400	—	300
	575	—	400	—	300
7	230	—	300	—	—
	460	—	600	—	—
	575	—	600	—	—
8	230	—	450	—	—
	460	—	900	—	—
	575	—	900	—	—

Figure 12-2: Electrical ratings for ac magnetic contactors and starters. *(Cont.)*

load, a motor draws excessive current that causes overheating. Since motor winding insulation deteriorates when subjected to overheating, there are established limits on motor operating temperatures. To protect a motor from overheating, overload relays are employed on a motor control to limit the amount of current drawn. This is overload protection, or running protection.

The ideal overload protection for a motor is an element with current-sensing properties very similar to the heating curve of the motor, which would act to open the motor circuit when full-load current is exceeded. The operation of the protective device should be such that the motor is allowed to carry harmless overloads, but is quickly removed from the line when an overload has persisted too long.

Fuses are not designed to provide overload protection. Their basic function is to protect against short circuits (overcurrent protection). Motors draw a high in-rush current when starting and conventional single-element fuses have no way of distinguishing between this temporary and harmless in-rush current and a damaging overload. Such fuses, chosen on the basis of motor full-load current, would blow every time the motor is started. On the other hand, if a fuse were chosen large enough to pass the starting or in-rush current, it would not protect the motor against small, harmful overloads that might occur later.

Dual-element or time-delay fuses can provide motor overload protection, but suffer the disadvantages of being nonrenewable and must be replaced.

The overload relay is the heart of motor protection. It has inverse trip-time characteristics, permitting it to hold in during the accelerating period (when in-rush current is drawn), yet providing protection on small overloads above the full-load current when the motor is running. Unlike dual-element fuses, overload relays are renewable and can withstand repeated trip and reset cycles without need of replacement. They cannot, however, take the place of overcurrent protective equipment.

Chapter 13

LIGHTING INSTALLATIONS

A simple lighting branch circuit requires two conductors to provide a continuous path for current flow. The usual lighting branch circuit operates at either 120 or 277 V; the white (grounded) circuit conductor is therefore connected to the neutral bus in the panelboard, while the black (ungrounded) circuit conductor is connected to an overcurrent protection device.

Lighting Control

Many lighting-control devices have been developed since Edison's first lamp, but the very basic devices include:

- Automatic timing devices for outdoor lighting
- Dimmers for adjusting the intensity of lamps in lighting fixtures
- Remote-control relays
- Common single-pole, 3- and 4-way wall switches

Switches

A wall switch, for our purposes, is a device used on branch circuits to control lighting. Switches fall into the following basic categories:

- Snap-action switches
- Mercury switches
- Quiet switches

Snap-action switches: A single-pole snap-action switch consists of a device containing two stationary current-carrying elements, a moving current-carrying element, a toggle handle, a spring, and a housing. When the contacts are open, as in Figure 13-1, the circuit is "broken" and no current flows. When the moving element is closed, by manually flipping the toggle handle, the contacts complete the circuit and the lamp will be energized. See Figure 13-2.

Mercury switches: Mercury switches consist of a sealed capsule containing mercury, as illustrated in Figure 13-3. Inside the capsule are contacting surfaces "A" and "B" which may be part of the wall of the capsule. The switch is operated by means of a handle which moves the capsule.

As shown in Figure 13-3, the capsule is tilted so that the mercury "C" has collected at one end of the capsule. Here, it bridges two contact points, "A" and "B" to complete the circuit and light the lamp. However, if the capsule is tilted the opposite way, the circuit between contacts "A" and "B" will not be completed, and the lamp will be de-energized (be turned off). Mercury switches offer the ultimate in silent operation and are recommended where the "clicking" of a light switch may be annoying.

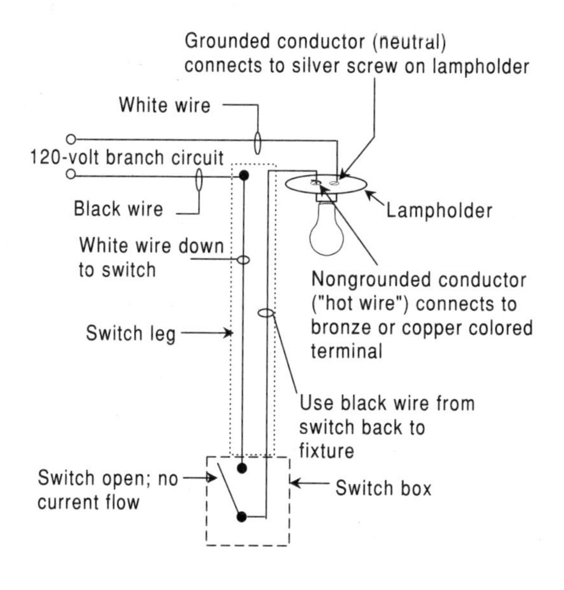

Figure 13-1: Switch in open (OFF) position.

Quiet switches: The quiet switch is a compromise between the snap-action switch and the mercury switch. Its operation is much quieter than the snap-action switch, yet it is not as expensive as the mercury switch.

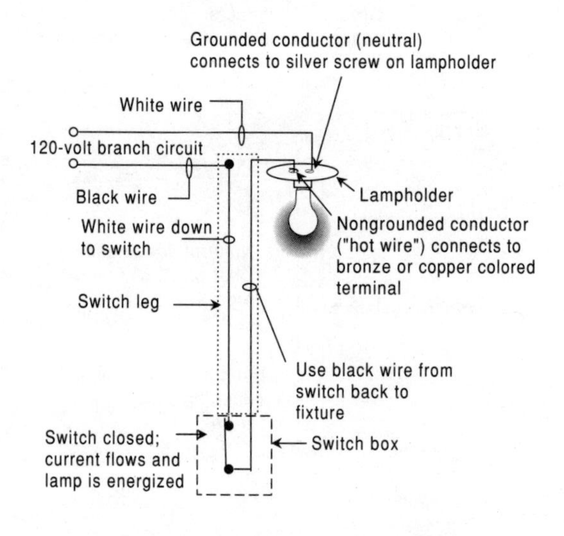

Figure 13-2: Switch in closed (ON) position.

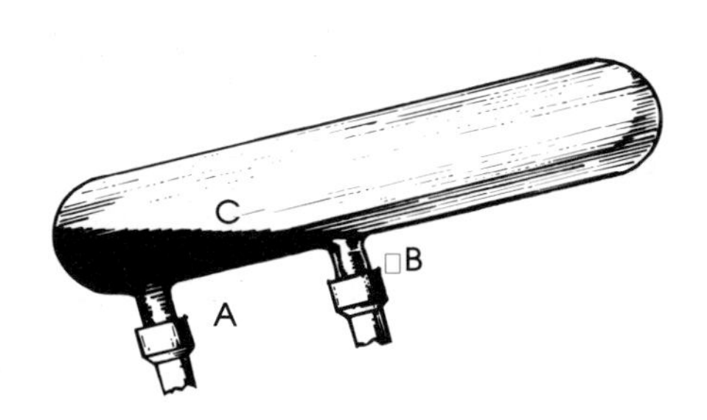

Figure 13-3: Principles of mercury-switch operation.

The quiet switch consists of a stationary contact and a moving contact that are close together when the switch is open. Only a short, gentle movement is applied to open and close the switch, producing very little noise. This type of switch may be used only on alternating current, since the arc will not extinguish on direct current.

The quiet switch (Figure 13-4) is the most commonly used switch for modern lighting practice. These switches are common for loads from 10 to 20 amperes, in single-pole, three-way, four-way, or other configurations.

Many other types of switches are available for lighting control. One type of switch used mainly in residential occupancies is the door-actuated type which is generally installed in the door jamb of a closet to control a light inside the closet. When the

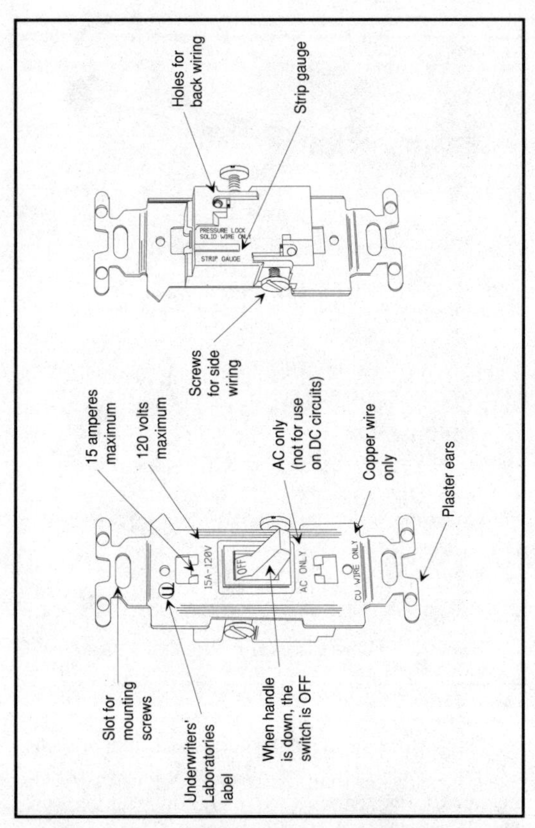

Figure 13-4: Characteristics of a single-pole quiet switch.

door is open, the light comes on; when the door is closed, the light goes out. Refrigerator and oven lights are usually controlled by door switches.

The Despard switch is another special switch. Due to its small size, up to three may be mounted in a standard single-gang switch box. Weatherproof switches are made for outdoor use. Combination switch-indicator light assemblies are also available for use where the light cannot be seen from the switch location, such as an attic or garage. Switches are also made with small neon lamps in the handle that light when the switch is off. These low-current-consuming lamps make the switches easy to find in the dark.

Three-Way Switches

Three-way switches are used to control one or more lamps from two different locations, such as at the top and bottom of stairways, in a room that has two entrances, etc. A typical three-way switch is shown in Figure 13-5. Note that there are no ON/OFF markings on the handle. Furthermore, a three-way switch has three terminals. The single terminal at one end of the switch is called the common (sometimes hinge point). This terminal is easily identified because it is darker than the two other terminals. The feeder ("hot" wire) or switch leg is always connected to the common dark or black terminal. The two remaining terminals are called traveler terminals.

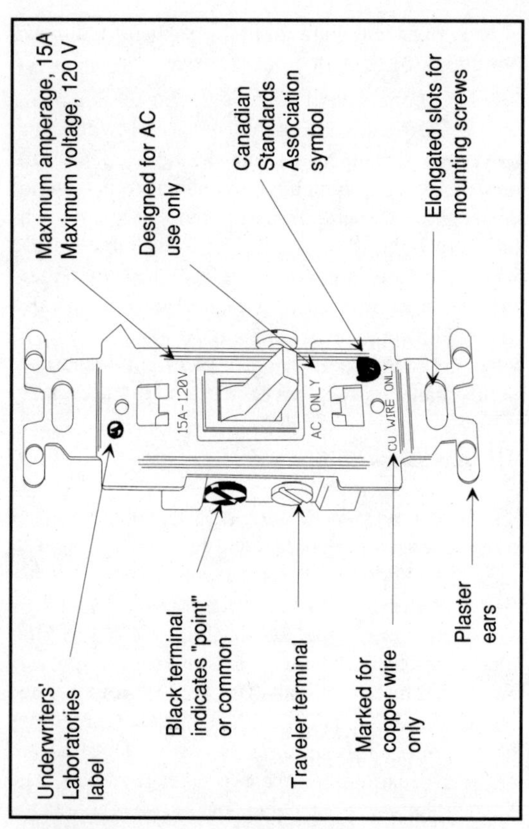

Figure 13-5: Characteristics of a three-way quiet switch.

These terminals are used to connect three-way switches together.

The connection of three-way switches is shown in Figure 13-6. By means of the two three-way switches, it is possible to control the lamp from two locations. By tracing the circuit, it may be seen how these three-way switches operate.

A 120-volt circuit emerges from the left side of the drawing. The white or neutral wire connects directly to the neutral terminal of the lamp. The "hot" wire carries current, in the direction of the arrows, to the common terminal of the three-way switch on the left. Since the handle is in the up position, the current continues to the top traveler terminal and is carried by this traveler to the other three-way switch. Note that the handle is also in the up position on this switch; this picks up the current flow and carries it to the common point, which, in turn, continues on to the ungrounded terminal of the lamp to make a complete circuit. The lamp is energized.

Moving the handle to a different position on either three-way switch will break the circuit, which in turn, de-energizes the lamp. For example, let's say a person leaves the room at the point of the three-way switch on the left. The switch handle is flipped down, giving a condition as shown in Figure 13-7. Note that the current flow is now directed to the bottom traveler

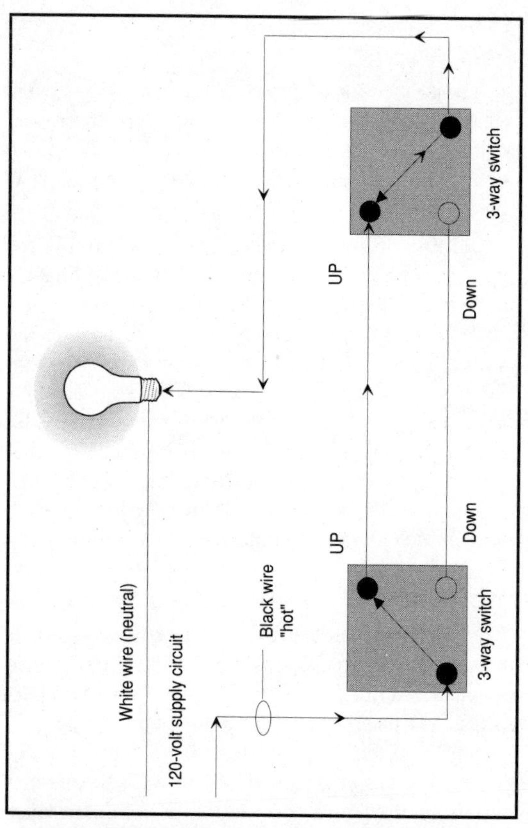

Figure 13-6: Three-way switches in ON position; both handles up.

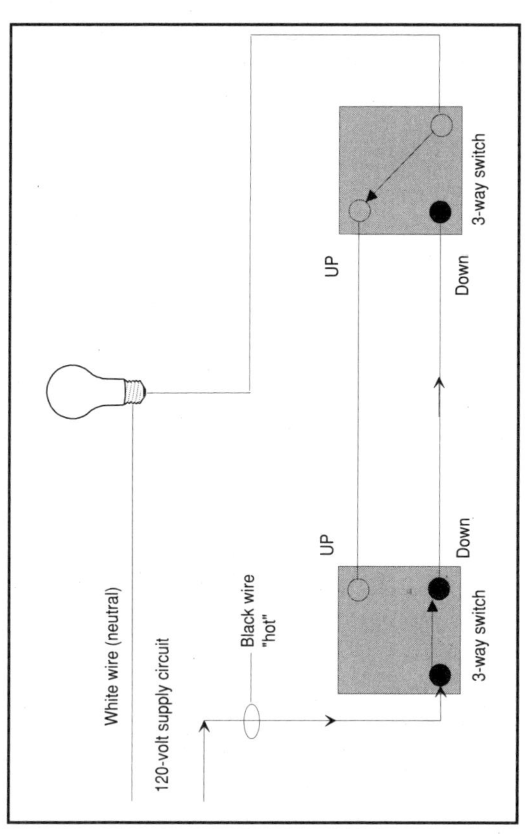

Figure 13-7: Three-way switches in OFF position; one handle down, one handle up.

terminal, but since the handle of the three-way switch on the right is still in the up position, no current will flow to the lamp.

Another person enters the room at the location of the three-way switch on the right. The handle is flipped downward which gives the condition as shown in Figure 13-8. This change provides a complete circuit to the lamp which causes it to be energized. In this example, current flow is on the bottom traveler. Again, changing the position of the switch handle (pivot point) on either three-way switch will de-energize the lamp.

In actual practice, the exact wiring of the two three-way switches to control the operation of a lamp will be slightly different than the routing shown in these three diagrams. There are several ways that two three-way switches may be connected. One solution is shown in Figure 13-9. Here, 14/2 w/ground NM cable (Romex) is fed to the three-way switch on the left. The black or "hot" conductor is connected to the common terminal on the switch, while the white or neutral conductor is spliced to the white conductor of 14/3 w/ground NM cable leaving the switch. This 3-wire cable is necessary to carry the two travelers plus the neutral to the three-way switch on the right. At this point, the black and red wires connect to the two traveler terminals, respectively. The white or neutral wire is again spliced — this time to the white wire of

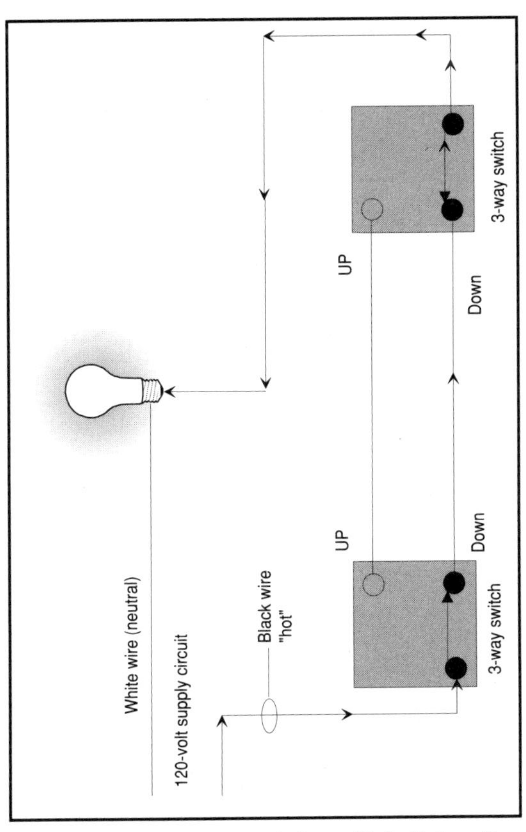

Figure 13-8: Three-way switches with both handles down; light is energized.

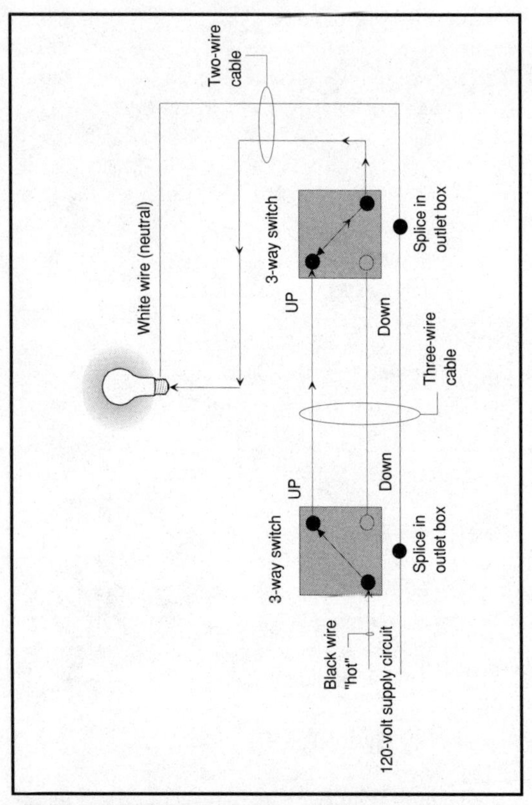

Figure 13-9: One way to connect a pair of three-way switches to control one lighting fixture or a group of lighting fixtures.

another 14/2 w/ground NM cable. The neutral wire is never connected to the switch itself. The black wire of the 14/2 w/ground NM cable connects to the common terminal on the three-way switch. This cable — carrying the "hot" and neutral conductors — is routed to the lighting fixture outlet for connection to the fixture.

Another solution is to feed the lighting-fixture outlet with two-wire cable. Run another two-wire cable — carrying the "hot" and neutral conductors — to one of the three-way switches. A three-wire cable is pulled between the two three-way switches, and then another two-wire cable is routed from the other three-way switch to the lighting-fixture outlet.

Some electricians use a short-cut method by eliminating one of the two-wire cables in the preceding method. Rather, a two-wire cable is run from the lighting fixture outlet to one three-way switch. Three-wire cable is pulled between the two three-way switches — two of the wires for travelers and the third for the common-point return. This method is shown in Figure 13-10, but should not be used with a metallic conduit system.

Four-Way Switches

Two three-way switches may be used in conjunction with any number of four-way switches to control

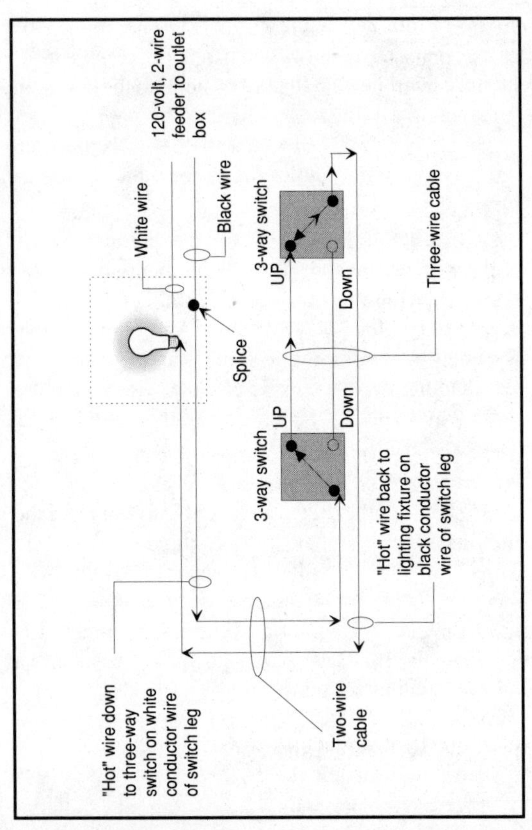

Figure 13-10: Alternate method of connecting three-way switches.

lighting from any number of positions. When connected correctly, the actuation of any one of these switches will change the operating conditions of the lamp(s); that is, either turn them OFF or ON.

Figure 13-11 shows how a four-way switch may be used in conjunction with two three-way switches to control a lamp from three locations. In this example, note that the "hot" wire is connected to the common terminal on the three-way switch on the left. Current then travels to the top traveler terminal and continues on the top traveler conductor to the four-way switch. Since the handle is up on the four-way switch, current flows through the top terminals of the switch and onto the traveler conductor going to the other three-way switch. Again, the switch is in the up position. Therefore, current is carried from the top traveler terminal to the common terminal and on to the lighting fixture to energize it. Under this condition, if the position of any one of the three switch handles is changed, the circuit will be broken and no current will flow to the lamp.

For example, let's assume that the four-way switch is flipped downward. The circuit will now appear as shown in Figure 13-12, and the light will be out. The light would also go out if the position of either the right or left three-way switch handle were changed.

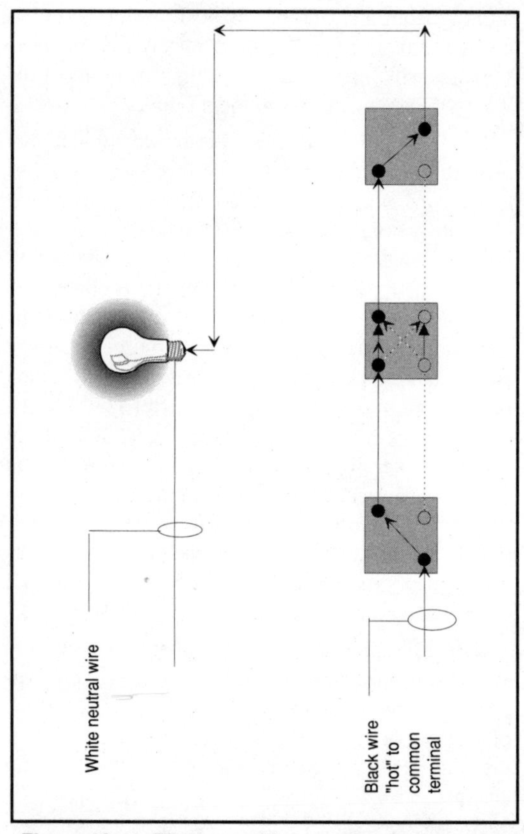

Figure 13-11: Three- and four-way switches used in combination; light is ON.

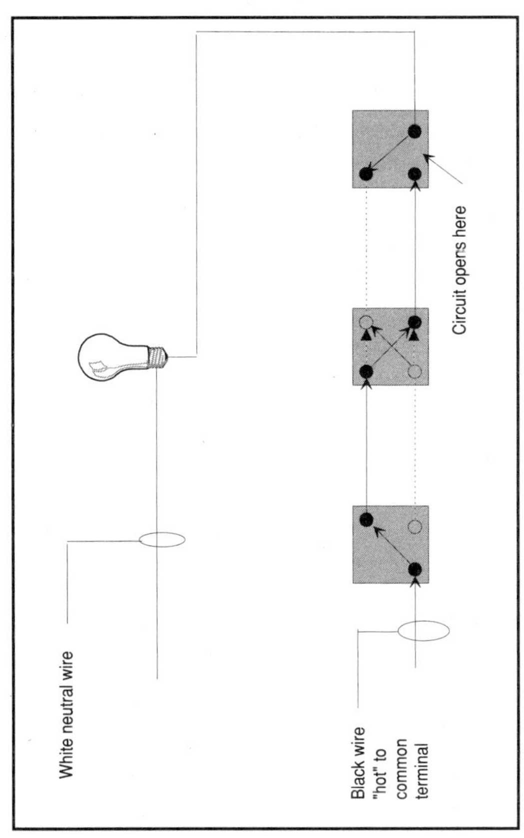

Figure 13-12: Three- and four-way switches used in combination; light is OFF.

Remember, any number of four-way switches may be used in combination with two three-way switches, but two three-way switches are always necessary for the correct operation of one or more four-way switches.

Photoelectric Switches

The chief application of the photoswitch is to control outdoor lighting, especially the "dusk-to-dawn" lights found in suburban areas.

Relays

Next to wall switches, relays play the most important part in the control of light. An electric or electronic relay is a device whereby an electric current causes the opening or closing of one or more pairs of contacts. These contacts are usually capable of controlling much more power than is necessary to operate the relay itself. This is one of the main advantages of relays.

Dimmers

Dimming a lighting system provides control of the quantity of illumination. It may be done to create various atmosheres and moods or to blend certain lights with others for various lighting effects.

A common application of this is in an auditorium used for plays, movies, and similar functions. To avoid shock or surprise when the lights are suddently turned on or off, or to avoid discomfort to the dark-adapted eye when the lights are suddenly turned on, dimmers are used to make this transition gradual.

Index

A

Advisory rules (NEC), 21
Aluminum connections, 316-318
Apprentice electricians
- employment, 14
- hours in training, 15
- knowledge of NEC, 19
- licensed, 14
- safety, 15
- training of, 15
- work of, 15

Apprenticeship
- advantages of, 13
- BAT programs, 12
- history of, 10
- modern training, 11, 13

Architect
- relationship with consultants, 8
- relationship with owner, 5
- working drawings, 1, 5, 9

Asbestos standards, 80-82

B

Back-to-back bends, 131-133
Bad habits in use of tools, 55-57
Bending conduit
- back-to-back bends, 131-133
- concentric bending, 162-165
- geometry of, 120-123
- leveling conduit, 173
- making 90-degree bends, 124-131
- mechanical offsets, 143-146
- mechanical stub-ups, 142, 143
- offsets, 133-136, 165-170
- PVC conduit (*see* PVC conduit installation)
- saddle bends, 137-138, 170-172
- tools used, 118, 119, 138-142, 146, 147, 150, 151, 173

Box connections, 226
Boxes
- cutout, 224-226
- installation of, 193
- junction, 193, 210, 213
- NEC requirements for sizing, 198-209
- outlet, 193, 198
- pull, 193, 210, 213
- removing knockouts, 197
- sizing of pull and junction, 213-223
- types of, 194-196

Boxes *(Cont.)*
where installed, 193, 210
Building construction
history of, 1
largest industry, 1
manuals, 17-18
organization and structure, 16
policies and procedures, 16
Bureau of Apprenticeship and Training (BAT), 12
Bushings, 227-230
Busways
types of, 247
use of, 252

C

Cabinets
NEC installation requirements, 224
Cable pull operations, 263
Cable systems
flat-cable assemblies (FC), 237
flat conductor cable (FCC), 237
metal-clad (MC), 239
mineral-insulated metal-sheathed (MI), 238
power and control tray (Type TC), 239
service-entrance (SE), 235, 236
Cable Systems *(Cont.)*
type AC (BX), 234
type NM (Romex), 232, 233
underground feeder (UF), 237, 235
underground service-entrance (USE), 237
Cable trays, 252
Canadian Standards Association (CSA), 277
Common power supplies, 288
Company procedures, 16
Concealed wiring systems
where used, 232
Concentric bends, 162-165
Conductor terminations
definition of, 307
Conductors in conduit
basket grips, 258
installation of, 255, 259-260
pulling location, 261
vacuum fish-tape systems, 256-257
wire caddies, 259
Conduit and raceway tools, 69
Conduit fill requirements, 102-105
Conduit, installation of
bending conduit, 118
cutting, 181-183, 240-241

Conduit, installation of (Cont.)
  factory elbows, 109
  joining, 186-190
  on-the-job bends, 110
  NEC requirements, 110-111
  PVC (*see* PVC conduit installations)
  reaming, 184186
  types of bends, 113, 116
Cutout boxes, *see* Boxes, cutout

D

Definitions in NEC, 41-42
Dimmers, 360
Distribution equipment, selection of, 292-294
Divisions of NEC, 24-27

E

Electric
  drills, 70
  hammers, 71
  shock, 94-97
Electric conduit benders
  making bends, 147-149
  operation of, 146
  speed benders, 150
Electric motors
  controls, 331
  importance of, 321
  nameplates, 332
Electric motors *(Cont.)*
  NEC requirements for installation, 323
  overload protection, 334, 338-340
  three-phase type, 321
Electric service
  common power supplies, 288
  grounding, 295
  parts of, 287
  selection of distribution equipment, 292-294
  service disconnecting means, 288
Electrical connections
  aluminum, 316
  basic requirements for, 307
  stripping and cleaning, 309
  wire nuts, 318-320
  wires under 600 V, 310
Electrical contractors covered by OSHA, 77
Electrical joints, 307-308
Electrical metallic tubing (EMT)
  cutting, 243
  installation of, 242
  use of, 242
Electrician
  apprentice, 14
  primary function, 9
  responsibilities, 9-10
  safety, *see* OSHA

Electrician *(Cont.)*
tools of, *see* Tools of the trade
tricks of the trade, 173
Employability, 99
Employee rights, 85-86
Enclosures, panelboard, 302

F

Fitzgerald Act, 12
Flexible metal conduit
installation of, 244
types of, 243
use of, 244
Four-way switches, 355-360

G

General classification of tools, 65-69
Geometry of bending conduit, 120-124
Grounding, 295-296

H

Hand tools, list of, 52
Hydraulic conduit benders
concentric bending, 162-165
making bends, 152-161
types of, 151

I

Identifying conductors
Identifying conductors *(Cont.)*
changing colors, 255
color coding, 254
NEC requirements, 253

J

Joining conduit, 186-190
Junction boxes
NEC installation requirements, 210-212
sizing of, 213

K

Knockout punch, 227
Knockout, removal of, 197
Knockouts in panelboard, 300

L

Lighting control
basic devices, 341
switches, 342
Locknuts, 227, 228

M

Making 90-degree bends, 124-131
Mandatory rules (NEC), 20
Manuals
company, 17
employee, 17
job, 18

Manuals *(Cont.)*
office, 18
Mechanical benders, 138-142
Mechanical offsets, 143-145
Mechanical stub-ups, 142-143
Mercury switches, 342-343
Miscellaneous power tools, 73
Motor controllers
function of, 331
nameplates, 332-333
Motor nameplates, 332-333

N

National Electrical Code (NEC)
address of, 14
bullets, 23
change bar, 14
definitions, 41-42
divisions of, 24-27
equipment for general use, 28-29
explanatory material, 21
extracted text, 23
language of, 21
meaning of different type, 23-24
practical application, 37-41
published by, 14
rules, 20-21
special conditions, 34
NEC *(Cont.)*
special equipment, 32-34
special occupancies, 30-32
terminology, 20-24
use of, 14, 35-36
wiring design and protection, 26
wiring methods and materials, 26-28
National Electrical Manufacturers Association (NEMA), 46
National Fire Protection Association, 14, 47
Nationally Recognized Testing Laboratory (NRTL), 45

O

Occupational injury and illness, 83-84
Occupational Safety and Health Act, The Federal, (OSHA)
asbestos standards, 80-82
available from, 76
definition of, 75
electrical contractors covered by, 77
employees rights, 85-87
goals of, 76
not covered by, 78
occupational injury and illness, 83-84
on-site reminders, 84

Occupational Safety and Health Act, The Federal, (OSHA) *(Cont.)*
recordkeeping, 82
safe motor vehicle standards, 80
scaffolds, 89-92
signs, 87-89
Offsets, 133-136, 165
the electrical worker and
OSHA safety standards, 78-79
On-site reminders (OSHA), 84
One-shot bending, 152-153
Open-wiring system
where used, 231
Outlet boxes, *see* Boxes, outlet
Overload protection, 334, 338-340

P

Panelboards
connections, 301
consists of, 297
enclosures, 302
general classifications, 297
knockouts, 300
mounting classifications, 297
Photoelectric switches, 360
Planning wire pulling installation, 259
Portable band saw, 73
Practical application of using NEC, 37-41
Proper care and use of tools, 62-64
Pull boxes
sizing of, 213-223
PVC conduit installations
bending, 176-180
cutting, 174
making joints, 175
where used, 174
Pulling location of conductors, 261

Q

Quiet switches, 344-345

R

Raceways
bending conduit, 118
consists of, 240
definition of, 101
function of, 101, 240
installing conduit, 109-113
NEC requirements for, 102
number of conductors permitted in, 102-105
tricks of the trade, 173
types of, 101
types of bends (*see* Bending conduit)
PVC (see PVC conduit installation)
Reaming conduit, 184-186

Receptacles
characteristics of, 270, 276
conductor markings, 277
definition of, 269
markings on, 277, 278
testing laboratory label, 277
Recordkeeping for OSHA, 82, 84
Relays, 360
Replacement of tools and equipment, 64

S

Saddle bends, 137-138
Safe motor vehicle standards, 80
Safety, *see* OSHA
Safety and your income, 99
Safety switches, 303-306
Scaffolds
inspection of, 91
types of, 90
use of, 89
Segment bends (90-degree), 153-162
Service disconnecting means, 288
Shock, electric, 94-96
Signs
caution, 88
danger, 87
safety first signs, 89
Snap-action switches, 342
Special conditions (NEC), 34
Special equipment (NEC), 32-34
Special occupancies (NEC), 30-32
Speed benders, 150
Stripping and cleaning conductors, 309
Surface metal molding
installation of, 245-246
use of, 244
Switches
common switch terms, 280, 281
four-way, 355
identification, 282
mercury switches, 342, 343
NEC requirements for, 279
photoelectric, 360
purpose of, 279
quiet, 344
snap-action, 342
three-way, 347
types of, 279-280

T

Terminology of NEC, 20-24
Testing Laboratories
list of, 44-47, 277
role of, 48
Three-phase motors
basic types, 321
motor connections, 324
operating principles, 321

Three-phase motors *(Cont.)*
  synchronous speed, 324
Three-way switches,
  347-354
Tools of the trade
  bad habits in use of, 55-57
  care and use, 62-64
  classification of, 65-69
  electrician's
    responsibility, 51
  knowing the tools, 53
  list of hand tools, 52
  replacement of, 64
  specific types of, 69-73
Trade workers, 14

U

Underfloor raceway, 253
Underwriters' Laboratories,
  Inc. (UL), 277
Using the NEC, 35-36

V

Vacuum cleaners, 71

W

Wire nuts, 318-320
Wireways
  definition of, 246
  installation of, 247
Wiring design and
  protection (NEC), 26
Wiring devices
  definition of, 269
  list of, 269
Wiring methods, 231
Wiring methods and
  materials (NEC), 26
Working drawings
  architect, 1
  cross sections, 5
  elevations, 1
  floor plans, 5
  large-scale renderings, 5
  plot plan, 1